# Green, Reliable and Viable

# Green, Reliable and Viable

## Perspectives on India's shift towards low-carbon energy

**Editors**

Ajay Mathur
Adair Turner
Noëmie Leprince-Ringuet

THE ENERGY AND RESOURCES INSTITUTE
*Creating Innovative Solutions for a Sustainable Future*

CRC Press
Taylor & Francis Group

CRC Press
Taylor & Francis Group
6000 Broken Sound Parkway NW, Suite 300
Boca Raton, FL 33487-2742



CRC Press is an imprint of the Taylor & Francis Group, an informa business



Printed on acid-free paper

International Standard Book Number-13: 978-0-367-27308-8 (Hardback)



Library of Congress Cataloging in Publication Data
A catalog record has been requested

Visit the Taylor & Francis Web site at
http://www.taylorandfrancis.com

and the CRC Press Web site at
http://www.crcpress.com

# Table of Contents

# Foreword

*By Ajay Mathur*

In the 1920s, when electricity was being extended to all households of the United States, one Tennessee farmer commented on this transformation, saying: "The greatest thing on earth is to have the love of God in your heart, the next greatest thing is to have electricity in your home". Throughout history, the provision of affordable and reliable electricity has been one of the keys to transforming lives for the better.

It is unquestionable that India's energy demand and consumption will continue to rise in the decades to come. India is currently among the world's fastest growing emerging economies, and the world's second most populous country. Thus, rapid growth in energy consumption will occur to fuel industrialisation, urbanisation, infrastructure development, and rising incomes. Such energy demand growth is essential for development and to the goal of bringing modern energy services to the millions of Indians still lacking them today. And this is India's prerogative: meaningful, reliable and affordable energy access for all.

India has taken remarkable strides in this regard in recent decades, with costs of renewable energy (RE) falling precipitously, electricity shortages declining, achieving electrification of every village in India in April 2018, and connectivity nearing 100% of households. The regulatory structure in place is geared towards making electricity affordable for all, with the Saubhagya Scheme for example, providing last mile connectivity and free electricity connections to all rural households and Below Poverty Line urban households. However, there is still a long way to go. India's per capita electricity consumption is just one-third of the world average. Electricity generation is also the largest contributor to India's energy related carbon dioxide ($CO_2$) emissions, and a major contributor to local air pollution and other environmental problems.

Cognisant of this fact, and in a push to reconcile the conflicting imperatives of development and environment, the Government of India has shown that it is committed to enhancing the 'greenness' of its energy footprint. In fact, India is deploying one of the world's most ambitious RE programmes, targeting 175 GW RE installed capacity by 2022. When India announced this target in 2015, some judged it as unrealistic, because back then, the share of RE in the total electricity generation was low. But solar installed capacity in India has increased by about 9 times from 2.63 GW to 23.28 GW between March 2014 and August 2018. The National Electricity Plan of India 2018 now foresees 275 GW RE by 2027.

This makes India a key country for global energy transitions. By implementing its RE targets successfully, India can act as a pioneer in leapfrogging to a new paradigm of development and industrialisation. Given its scale, circumstances, rapid growth and commitment to sustainable development, India's energy transitions present tremendous opportunities, but of course, this comes with its set of challenges.

This is a learning process for India. The first phase of our energy policy focused on creating the conditions to bring affordable and reliable energy to all, and this has been achieved spectacularly. With these strong foundations in place, we must now look to the next phase of our energy policy as future-looking, aiming to balance conventional and alternate sources of energy along with energy efficiency, going beyond the energy access objective and pursuing the goal of energy independence. This will only be possible with a much higher share of RE in our energy mix, and this transition must be carefully planned and prepared for, starting now.

For this to happen, we must understand that this is not one transition, but many transitions happening and coming together. A plethora of key actions, policies and finance mechanisms, changing institutions, markets, business models, infrastructure and behaviours, can make it feasible for India to meet its rapidly rising energy demand with an increasing share of RE, high standards for affordability and reliability, and minimal disruption in all the involved sectors.

This is not to say that coal will not remain the dominant energy supplier over the course of the next decade; but India needs a strategy going beyond the 2022 short-term milestone, which is not an end in itself. This should be a stepping stone towards a longer-term, affordable and reliable energy system that does not look anything like today's. Unleashing the full potential of transitions will require significant structural and institutional changes, regulatory support and interventions, long-term policies and a critical mass of key stakeholders supporting this orientation.

This book of thought-leadership exemplifies the support that is required for these transitions to occur. As shows this book, energy transitions concern every sector, and all stakeholders, from industry, to finance to government. Stakeholders must work together to bring to the table the cooperation that forms the core of the integrated approach that will drive these energy transitions. I much welcome the foundation laid by this book, on which I hope we can build momentum, to lay out the transformational paths of change that our country needs to embark on.

*Dr Ajay Mathur is Director General of The Energy and Resources Institute*

## Bibliography

Gordan, J., R., (2016). *The Rise and Fall of American Growth: The U.S. Standard of Living Since the Civil War.* Princeton, New Jersey: Princeton University Press.

Green, Reliable and Viable

# Preface

*By Adair Turner*

I have often said that the resource-intensive practices of yesterday can no longer sustain the world. The future of our planet hinges on timely transitions to resource-use efficiency across ecosystems of people, products, and processes. Aligned with this thought, the India spin-off of the Energy Transitions Commission (ETC India) was launched in February 2018 to spearhead the nation's transition to low-carbon energy systems. As we reach the first anniversary of ETC India, it is time to pause and assess the challenges, learnings, and the way ahead.

India, which is undertaking the world's largest household electrification programme, is faced with the dual challenge of achieving universal access to reliable power while reducing its reliance on fossil fuels. Moreover, the country has committed, under its Nationally Determined Contributions, to reduce the emissions intensity of its gross domestic product by 33% to 35% by 2030 from 2005 levels, and to ensure at least 40% non-fossil fuel-based electricity generation capacity by 2030. This is a noteworthy resolve for a nation that is investing heavily in energy systems, mobility, industrialisation, infrastructure, housing, and urban facilities amongst others.

I firmly believe that the road to achieving these goals goes through a low-carbon economy. The importance of this transition becomes even more pronounced when analysed in the Indian context. The country is making giant strides in transforming its energy mix with significant additions in renewable energy capacity, enabling policies in its grid integration, and large-scale measures in the field of energy efficiency. While these targets are noteworthy in themselves, it is equally important to evaluate them against India's changing energy landscape and ever-increasing demand for energy. The big picture, therefore, highlights the imminent need to accelerate the transition to low-carbon energy systems.

This book includes valued perspectives from key stakeholders in this transition.

Experts and practitioners from the power, mobility, agriculture and energy efficiency sectors, amongst others, have shared their outlook on how to accelerate the ongoing shifts to green energy systems; on the next steps required to make those green shifts reliable; and further, to make them viable for end-use sectors and consumers. They have provided their perspectives on the policy, corporate and financial aspects that will enable these transitions. The overarching message reaffirms that the Indian energy sector of the future will be noticeably different from its current version. However, to enable this transformation, there is a need for new technology, business models and policies that are specifically tailored to the Indian context. There is also a need for regulatory frameworks to deal with institutions and infrastructure that will be supplanted.

For instance, stakeholders argue for digitalisation of the electricity grid, to drive cost-efficiency while leveraging automation of end-to-end processes in power plants (from coal yard management to ash utilisation). Or an entire off-shore platform to be managed remotely, efficiently and with fewer people. Despite being a path-breaking transition in ways more than one, such a step would need to be supplemented with adequate frameworks dealing with the eliminated processes, resources and structures.

Another enabler of low-carbon energy transition is energy storage technology, which has been emphasised in this book as a necessity to ensure flexibility of the electricity system. In the chapter on energy flexibility, the author argues that energy storage (along with Demand Reduction Measures and utilisation of existing thermal plant capacity) is a prerequisite for India's transition to renewables. It reduces the need for expensive backup capacities, by storing part of the solar output during the day and releasing it back to the power system in the evening. Developing a timely and integrated approach to enable solutions such as energy storage, interconnected grid, along with demand-side measures, effective markets and cooperation between state and central authorities are crucial for India. Similarly, blockchain technology is projected to play an important role in connecting energy consumers (even in remote off-grid rural areas) to investors, producers, and grid operators.

In the chapter on mobility, the author argues that the future of mobility in India needs to be shared, convenient, connected, reliable and electric. With a growing demand for mobility in India, the potential for scaled adoption of electric vehicles across different customer segments, including passenger as well as freight movement is significant. This, again, builds a strong case for transition to clean and low-carbon transport systems – the dividends of which will not only provide an impetus to climate action but also improve mobility standards and savings for end-consumers.

In the chapter on energy efficiency, the author advocates future-oriented technologies such as trigeneration, electric mobility and smart meters while calling them the 'starting point' for energy transitions. According to him, far-reaching results will be generated through business models that are tailor-made for the Indian context. Similarly, the chapter on *Managing India's Renewables Target* lists down five key steps that India must take to enable this transition. Hybridisation of solar and wind energy, development of robust ancillary markets to support the grid infrastructure, and increased investments in high-voltage transmission lines are the top three.

Further, software solutions to optimise grid-level operations as well as consumer-level behaviour, and battery storage have been advocated as the next key steps. As the cost of battery storage falls, it will become increasingly viable to manage intermittency in the grid integration of renewable energy. Aligned with the thought of investing in future-oriented technology, the chapter on *Enhancing the Green Footprint of the Electricity Transition in India*, states that in the short-term, the country needs to consider technological interventions which also reduce air pollution and water-usage in the power sector. However, these need to be woven into the long-term strategy for a cleaner power system based on renewables.

Coming to the agriculture sector – the backbone of the Indian economy – experts believe that innovative approaches building towards a more sustainable and equitable scenario will enable the low-carbon transition. Comprehensive efforts that facilitate lower water-intensive cropping patterns and extensive soil-water conservation coupled with rainwater harvesting schemes are critical to reduce the energy consumption in agriculture. Decentralised renewable energy solutions for agricultural pumping requirements will prove to be a game changer in this transition. Strong interlinkages between food, water, and energy, therefore, need to be recognised and analysed in an integrated manner to ensure a smooth and swift transition to low-carbon systems.

We know that the fight against climate change is an uphill task, but it is not an impossible one. The global community also recognises that the developing world's energy demand and consumption will rise during its journey so as to enable a quality of life that is at par with the developed world. Nonetheless, it is still possible to achieve these goals while containing our carbon footprint. The answer lies in transitioning to a low-carbon global energy system.

According to ETC's *Better Energy, Greater Prosperity* report released in April 2017, it is technically and economically feasible to grow economies and provide affordable, reliable, and clean energy for all while meeting the Paris objective of limiting global warming to well below 2°C. The key is to accelerate energy productivity and decarbonisation of the electricity sector, move towards increased electrification in homes, mobility and industry, and start introducing zero-carbon technologies in the so-called 'harder-to-abate' sectors, such as cement and steel, and trucking and aviation.

With multiple steps in the right direction, India has set the wheel in motion. It is now imperative that we add to this momentum. After all, the future of our planet hinges on the sustenance of this movement.

*Lord Adair Turner is Chair of the Energy Transitions Commission*

## Bibliography

The Energy Transitions Commission (2017). *Better Energy, Greater Prosperity: Achievable pathways to low-carbon energy systems.* Retrieved from http://energy-transitions.org/sites/default/files/BetterEnergy_fullReport_DIGITAL.PDF

# Acknowledgements

We would like to acknowledge the invaluable support from team Edelman, particularly Karnika Bahuguna, Anasuya Chatterjee and Vasudevan Rangarajan towards the production of this book. We would also like to thank the TERI team, Rishu Nigam and Thomas Spencer, for their valued editorial contributions. Finally, we would like to thank members of TERI Press, including Santosh Gautam, Sudeep Pawar, Abhas Mukherjee, Anushree Tiwari Sharma, Rajiv Sharma and Aman Sachdeva for their efforts to produce this book.

Green, Reliable and Viable

More than half of India's 2030 infrastructure is yet to be built. This is a challenge and an opportunity for the corporate sector to support India's decarbonisation

# Energy Transitions in India
## The Global Context

*by R R Rashmi*

Even before 'sustainable development' became the buzzword of environmental discourse, India had been making efforts to generate energy from alternative sources in order to meet its energy demand in a sustainable manner. This was guided by two primary considerations – one, of achieving energy security and second, of promoting energy access. A country which meets nearly 47% of its total primary energy demand with imported fossil fuels and whose demand is growing at the rate of 4% per annum, cannot be oblivious to potential disruptions in energy supply. It is critical for India to explore all sources of energy so as to devise a robust, stable, and secure energy system for its citizens. Limits in terms of natural resource endowments make it further necessary for Indian energy planners to devise an energy mix that is sustainable.

This policy yielded quick dividends. By 2008, India had established itself as a relatively strong leader in wind energy and announced its National Action Plan on Climate Change (NAPCC). India's NAPCC encapsulates missions for both solar energy and energy efficiency. While the goals envisaged under the two missions were a courageous departure from business-as-usual, the twin challenges of safeguarding the country against climate change, and bringing energy access to over 60% of its population obliged India to move towards a lower carbon society with greater urgency.

## Implications of the Paris Agreement

The challenge has become more intense in the wake of the Paris Agreement which requires every country to contribute to the global goal of mitigating climate change. Of course, the Agreement provides flexibility to each country to design its targets according to its national circumstances, but the obligation to do so is universal. India is bound by its Nationally Determined Contributions (NDCs) under the Agreement.

There are two specific implications of the Paris Agreement for the global emissions trajectory and national energy systems; both have strong consequences for India's energy story. First, the global emission pathways required for reaching the climate stabilisation goal of either 2°C or 1.5°C involve constrained global carbon space. The latest Intergovernmental Panel on Climate Change (IPCC) Special Report on 1.5°C (SR1.5) indicates that the global available carbon budget ranges from 570 to 770 gigatonnes of carbon dioxide ($CO_2$) for 66% and 50% probabilities of making the 1.5°C goal. If the goal is to be reached without overshooting, specific transformational energy technologies as well as carbon dioxide removal will be required. Seen against its NDCs, it is incumbent on India, as well as other major economies, to implement low-carbon strategies including appropriate policy and investment choices consistent with this budget.

Two, the Paris Agreement mandates each country to contribute to the global goal of mitigating climate change by reducing emissions in relative or absolute terms, according to national circumstances. This flexibility is contingent on a transparency obligation under which each country is accountable for its actions as per an agreed international protocol. In the absence of adequate actions for developing sustainable energy systems, the chances of a defaulting country being subjected to international criticism and called to order are fairly real. Both of these considerations make it vital for India to accelerate its energy transitions.

## Global Assessments for Possibilities

It is useful to look at the global assessment of the required transitions in energy systems before considering the possibilities for India. Both, the IPCC Fifth Assessment Report (AR5) and its recent SR1.5 contain assessments of the modelled emission pathways required for achieving the long-term goal.

The SR1.5 finds that 1.5 pathways 'would require rapid and far-reaching transitions in energy, land, urban and infrastructure (including transport and buildings), and industrial systems. These systems transitions are unprecedented in terms of scale, but not necessarily in terms of speed, and imply deep emissions reductions in all sectors, a wide portfolio of mitigation options and a significant upscaling of investments in those options.'

More specifically on energy systems, the SR1.5 indicates that 1.5°C consistent pathways would require faster electrification of energy end use, and higher shares of low-emission energy sources well before 2050. The report finds that the share of renewables for electricity must increase to 70%-85% by 2050, with the share of coal reduced to close to 0%-2% of electricity. The electricity share of energy demand in buildings must be about 55%-75% and the share of low-emission energy in the transport sector is to rise from less than 5% in 2020 to 35%-65% in 2050.

Although the options presented by the IPCC differ in degree of feasibility as countries have differing capacities, there is optimism that the political, economic, social and technical feasibility of solar energy, wind energy and electricity storage technologies will improve substantially soon. While achieving system transitions in electricity generation may not be easy, there is an emerging consensus that countries can lower the barriers and costs for investments through adoption of enabling policies, technological innovations, and adequate financing.

## The Challenge and Hope for India

The challenge for India in terms of low-carbon transitions is humungous. Energy emissions contribute 71% to its total $CO_2$ emissions and are expected to grow at around 3%-4% per annum till 2035. Due to continuing economic growth, India is expected to be the second largest contributor to the increase in global energy demand by 2035, accounting for 18% of the rise in global energy consumption. Given this, and India's NDCs which include moderation of emissions intensity of its gross domestic product (33%-35% below 2005 levels) and non-fossil fuel electricity generation (40%), by 2030, it is crucial that the Government of India and the corporate sector urgently internalise low-carbon options in their strategies to transform energy systems.

What gives hope is the fact that more than half of India's 2030 infrastructure is yet to be built, and improving energy access remains a government priority. This presents both a challenge and an opportunity for the corporate sector to support India's decarbonisation.

## Renewable Energy: Key to Energy Transitions in India

A shift to renewable energy (RE) is key to this strategy. India targets 175 GW of renewable capacity by 2022. Roughly 70 GW of renewable capacity has been installed and at least another 40 GW is under construction. If current capacity utilisation factors are considered, the targeted level could replace about 70 GW of coal capacity.

According to the latest National Electricity Plan of India (2018), the target can be scaled up to 275 GW by 2027. This will take the contribution of RE sources to the total energy demand to 24.4% in 2026-27. The Plan also suggests that no new coal-fired stations will be required during 2017-22, with current capacity and projected renewables capacity being sufficient to meet demand growth.

The central government offers favourable tariffs and subsidy programmes for installation of renewable power systems. States provide reduced transmission and distribution charges and follow competitive bidding in allocation of RE projects. An ambitious public procurement policy for solar power has lowered the price of RE in the last few years making it a cheaper source of power than coal (without considering the transmission and other attendant costs).

However, RE integration into the grid requires several technical challenges to be addressed. As RE availability is variable, dependence on RE requires support of storage systems which are not yet commercially viable at the required scale. In the absence of viable flexibility options, the baseload demand in the system has to be per force met with conventional sources of energy.

Peak demand management requires stable grid operations. Adequate infrastructure and trading institutions are therefore necessary for grid balancing, to manage the large variability of RE. Institutional capacity for improved inter-state cooperation has to be created for this to happen.

Towards this end, the Government of India has recently bid out a 100 GW solar tender, with an emphasis on battery storage and domestic solar manufacturing. This announcement follows plans for 8-10 GW of annual onshore wind installations, plus an ambitious 30 GW of offshore wind by 2030. The RE sector also creates jobs, with estimates being that the wind energy sector alone may employ 2 million people by 2022.

Transitions in the transport sector will also be required. The government intends to promote large-scale penetration of electric vehicles in the system but is, in the near-term, focused on creating infrastructure for charging and hybrid vehicles. The government has decided to advance implementation of vehicular fuel efficiency norms to 2020 to curb vehicular pollution and reduce carbon emissions.

## The Role of Industry Processes and Products

Energy transitions in India will not be complete unless they cover key industrial processes and products. Energy-intensive and harder-to-abate sectors such as cement, steel, aluminium, metals, buildings and transport, in particular, need to adopt decarbonising technologies of comparable efficiency and not-so-high investment risks. Innovations in such sectors include electrification of processes, cheaper hydrogen-based fuels, bioenergy-based applications, material efficiency, circularity and development of new materials.

There is increasing recognition that technological innovations and their commercial deployment at scale can make large scale transitions happen in such difficult sectors. Lowering of the investment risks is needed, and the development of a voluntary, climate related financial risk disclosure by companies can help accelerate the transitions. Use of explicit or implicit carbon pricing can aid this process. Financing models where grant-based or lower-cost public funds can be combined with private capital need to be developed in order to lower the risk for the investments and to keep the cost of power affordable to consumers.

It is clear that the energy transition in India entails change of great scale, complexity and uncertainty. It requires the buy-in of multiple stakeholders from industry, financial institutions, utilities and government. Moreover, it requires the establishment of an effective framework of credible long-term objectives – and near-term policies. Recent moves made by the government as well as the private sector in India give reasonable hope that such policies exist and that the private sector will find profitable opportunities to engage in attractive and sustainable markets for lower carbon investments.

*Mr R R Rashmi is Distinguished Fellow at The Energy and Resources Institute*

## Bibliography

Central Electricity Authority, MoP, GoI, (2018). *National Electricity Plan.* Retrieved from http://www.cea.nic.in/reports/committee/nep/nep_jan_2018.pdf

IPCC, (2018). Summary for Policymakers. In *Global warming of 1.5°C. An IPCC Special Report on the impacts of global warming of 1.5°C above pre-industrial levels and related global greenhouse gas emission pathways, in the context of strengthening the global response to the threat of climate change, sustainable development, and efforts to eradicate poverty.* World Meteorological Organization, Geneva, Switzerland.

IPCC, (2014). *Climate Change 2014: Synthesis Report. Contribution of Working Groups I, II and III to the Fifth Assessment Report of the Intergovernmental Panel on Climate Change.* IPCC, Geneva, Switzerland.

# SHIFTING TO GREEN ENERGY SYSTEMS

India needs to produce 275 GW from renewable energy by 2030 to meet its climate commitments

# Managing India's Renewables Target

*By Sumant Sinha*

India currently has a power generation capacity of 350 GW, generating 1.2 trillion units of electricity each year. This is equivalent to saying that each Indian citizen, on average, consumes 1,000 units of electricity annually. For comparison, China generates around 4,000 units per capita annually; the world average is around the same, and advanced economies generate upwards of 10,000 units of electricity per citizen each year. Therefore, it should be clear that India faces two immediate and urgent tasks if it aspires to grow ambitiously. First, it must upgrade and expand its chronically underperforming electricity sector; and second, it must clean the dirty power sources that currently account for two-thirds of the production value chain in this sector.

01

1. Per capita generation of electricity:
   - India ~ 1,000 units
   - China ~ 4,000 units
   - Advanced economies ~ 10,000 units or above

02

2. India's grid currently absorbs ~75 GW RE. To achieve 40% of total RE capacity by 2030, India needs an annual addition of ~16 GW over the next 12 years

## Demand for Electricity

Let's be clear, swelling of cities, electrification of villages, growth of new industries, expansion of existing ones, increased utilisation of air-conditioning, growth in the electric vehicle market, and other factors will contribute to the increase in electricity demand in India. Keeping in mind the economic aspirations of our people and our government, India should at least double the standard of living it provides to its average citizens in the next 10–12 years. To get there, electricity generation has to grow at a compounded growth rate of at least 6% annually, and that is not an unreasonable task – in the last decade, India's electricity generation has grown at a compounded rate of 5.7%. If we can simply maintain this level of growth, we would add another 1.2 trillion units of electricity to the grid and cater to India's latent demand for electricity.

The moot question is, how do we get there in a way that is clean? To double our existing thermal capacity from 220 GW to 440 GW would mean inviting millions of preventable cancer cases, lowering life expectancy on account of other pollution related ailments, and missing our emissions related targets. On that last point, we would not make our fair contribution to containing global temperature increase to less than 2°C from pre-industrial times to contain climate change. One knows the consequences of climate change. Consider what happens to India's National Capital Region every October – Delhi gets choked on severe pollution. Our air quality index skyrockets, in large parts due to surrounding coal power plants and vehicular emissions, to levels that exceed 20 times the most hazardous benchmark set by the World Health Organization. Can we afford to have 'October in Delhi' episodes around the country because we did not plan our electricity expansion efforts properly?

## Focus on Renewables

The Indian government recognises this. Renewables have cheaper levelised costs of energy – or lifetime operation costs – than thermal-based power, they are clean, they do not degrade the environment, and they do not spew greenhouse gases that accelerate climate change. To balance India's economic growth aspirations, local pollution crises, and to contribute to climate change containment commitments, the Indian government has aimed for renewables to account for roughly 40% of electric generation capacity by the year 2030. Assuming we produce and consume 2.5 trillion units of electricity annually by then, and do so with 40% renewables capacity, we would require roughly 275 GW of renewables by that year.

Presently, the country's grid absorbs circa 75 GW of renewables. To get to the 2030 target would require an annual capacity addition of around 16 GW of renewables a year for the next 12 years. The average rate of installation has so far been 6.5 GW a year. Thus, there is room to grow. It is hearty to note that in this current fiscal year, we have already installed 8 GW of renewable capacity. To ensure that we reach the 2030 target, however, we must pre-empt challenges and put in place plans to mitigate them. The foremost of these challenges is the fundamental issue of intermittency. The sun does not shine at night, and the wind does not blow round the clock. We cannot have lights go off every time there is a cloud cover over our solar panels or each time the wind slows down. But intermittency will become an acute problem when solar generation starts accounting for 20% of overall supply under today's grid conditions, and we are sometime away from that. That gives us enough room to prepare.

## Policy and Technology Interventions

Consider the following technology and public policy interventions:
The hybridisation of solar and wind energy: Power projects should be clubbed together so the combined electricity they inject to the grid is smoother in flow than it would have been if the power plants were injecting power independent of each other. Combined solar-wind projects have capacity utilisation factors of over 40%, closer to those of coal-fired power plants, practically eliminating the intermittency challenge. Similarly, floating solar, or solar panels on plastic floating structures placed on dam reservoirs, can couple very well with hydro power assets to provide a semblance on baseline power supply – hydro at night, and solar during the day. There is also a significant cost reduction in hybrid projects on account of sharing of transmission lines.

The development of robust ancillary markets: Ancillary services are backup services that smoothen out the variable nature of energy supply. Germany has close to 3% of spinning, storage, and other reserves to support its grid infrastructure; we do not have any. There are tremendous opportunities for improvement in this area alone.

Increased investments in high-voltage transmission lines to transport large amounts of energy over vast distances quickly and efficiently: It is apparent that Rajasthan, Gujarat, Maharashtra, Madhya Pradesh, Andhra Pradesh, Telangana, and Karnataka are leaders in renewable energy generation in India today. With a robust national transmission system, power produced in these states can be shipped to other power-deficient states like West Bengal, Odisha, Jharkhand, Bihar, Uttar Pradesh, Haryana, Kerala, and others. Transmission investments would be needed regardless of renewable capacity addition, given that power demand in India is suppressed and will increase in the coming decade.

Investments in software solutions to optimise grid-level operations as well as consumer-level behaviour: The creation of demand response programmes, for example, can prod industries to shift their loads to times during the day when more energy is available on the grid. This reduces peak demand. Demand response industries will not be cost burdens on renewables and are, in fact, proven business models in and of themselves. There are examples in developed countries where software companies serve as intermediaries and charge fees to both heavy and flexible energy users, like certain industries, and to distribution companies' (discoms), to manage their demand and supply situations, respectively, better.

Battery storage: As battery storage costs fall precipitously, they will become increasingly viable to manage intermittency. Grid operators can store electricity generated from renewable projects in large battery systems when demand is low and release that electricity into the grid when demand increases.

The need for discom reforms: Nearly 25% of electrons generated from power producers is lost en route to end-consumers, and much of this loss is on account of discoms. Discoms are unable to invest in upgrading their infrastructure and as a result face electricity losses due to line faults, line leakages, undersized and over-utilised transformers, electricity theft, and poorly monitored maintenance processes. They end up buying more electrons than they sell and this has negative ripple effects on the entire value chain. Discom privatisation in Delhi and Mumbai has shown remarkable progress on this issue where Aggregate, Techincal & Commercial losses of over 50% have been reduced to 15% in a single decade. Inefficiencies converted to revenue will enable discoms to make investments to build a smarter, digital grid, minimise losses, and better integrate renewable power into their districts.

Today renewables are more economical to install and commission than thermal-based power; they also enable growth and environmental conservation. While the government has taken many laudable steps to promote this technology, sensible, practical, and proven interventions can ensure that potential challenges are adequately managed. We should enable work towards a greener future for India that is possible.

*Mr Sumant Sinha is Chairman and Managing Director of ReNew Power Ltd.*

As urban India will increasingly present space limitations, there is a need to shift focus to decentralised small solar power installations in rural India

# Roadmap for a Solar Surge

*By Ajay Shankar*

In this decade India's solar energy programme has grown exponentially. Solar power capacity, which was less than 1,000 MW in 2010 when the National Solar Mission was launched, is now more than 26,000 MW. But India's success in solar power suffers a major distortion. The bulk of the solar power capacity has been created through large competitively selected solar projects which feed electricity into the grid. On the other hand, decentralised solar power capacity, including rooftop, is still less than 4,000 MW.

This is the exact opposite of what the lower cost option is. Decentralised rooftop solar power does not need any transmission investments. And there are no technical transmission and distribution losses. Further, capacity utilisation of transmission lines set up exclusively for solar projects is naturally lower, as solar power is generated only when the sun shines. This distortion will now start imposing avoidable costs, as the share of solar power in total generation begins to rise rapidly from the present negligible share of around 2%–3%.

The special dispensation of a zero transmission tariff for solar power will also start becoming unsustainable. At present the costs of transmission are borne through cross-subsidy from the transmission charges for the flow of conventional power. This will need to change and actual transmission costs will need to be explicitly paid.

## Exploring the Solar Potential of Rural India

This distortion needs urgent reversal. While efforts are being made to promote solar rooftop installations and some momentum is being seen with the generous net metering dispensation, there are limits in urban areas. Most of urban India is going through redevelopment. Older bungalows are giving way to multi-storied apartments where the available space for solar power panel installation becomes modest. In the slums as well as in congested old city centres, there are again space limitations. But the potential for decentralised small solar power installations in rural India is enormous. This is the segment that now needs focus.

For the large solar projects, competitive bidding has been working well. It has brought prices down and India has been able to take full advantage of the global decline in manufacturing costs of solar panels. But for small decentralised solar power installations in the less than 1 MW range, competitive bidding, project by project, is just not feasible. Only an attractive enough feed-in tariff would work. Germany, with not much sunshine, has 1.6 million solar photo voltaic generators and is a world leader in solar power with an installed capacity of 43,000 MW, thanks to a generous feed-in tariff regime.

A feed-in tariff regime for rural India would mean that the local power distribution company would announce that at a particular rural substation, or a distribution transformer, it would be willing to buy solar power at a feed-in tariff (preannounced price) for 20 years. This offer would be open on a first-come basis for two years, or until the technical capacity to absorb solar power at that point is reached. Purchase from a single supplier would be restricted to, say, 1 MW, so that a large number of small projects in the kilowatt range are promoted.

## The Investment Scenario

As solar power prices for the larger projects have come down to less than ₹3 per unit, a feed-in tariff of, say, ₹4.50 per unit should be attractive enough to get a surge in private investment to benefit from this opportunity. This would give higher incomes to farmers whose land is used, local persons who would install and maintain the solar panels and the providers of solar panel installation services. It would provide a fillip both for new business models – through RESCOs – as well as create a surge in employment. A feed - in tariff of, say, ₹4.50 would make this a cheaper source of additional power for the distribution company compared to electricity from conventional sources through normal power purchase agreements. At the point of feed-in within the rural grid, the marginal cost of conventional power after factoring in technical losses and the cost of transmission is usually over ₹6.50 per unit.

In fact, there is a good case to also shift from net metering to a feed-in tariff regime for solar power generation, including rooftop, within the premises of consumers. This would be attractive enough for those installing rooftop solar power and also for the distribution company, who find net metering financially too onerous. Such an attractive feed-in tariff regime would enable solar power to grow exponentially. Market forces would work wonders. No subsidies would be needed. All that the distribution companies would need to do would be to assess the quantity of electricity that they could comfortably take at a particular substation, and then announce that they would buy solar energy on a first-come, first-served basis up to this amount at the substation.

## Reliable Solar Power for All

Supply of solar power to rural substations in the daytime would also enable the distribution companies to supply reliable electricity to farmers for the whole day. This would make an enormous difference to the quality of life and work of farmers who are used to getting electricity for irrigation only during off-peak hours in the night. It should lead to better water-use efficiency, as in the

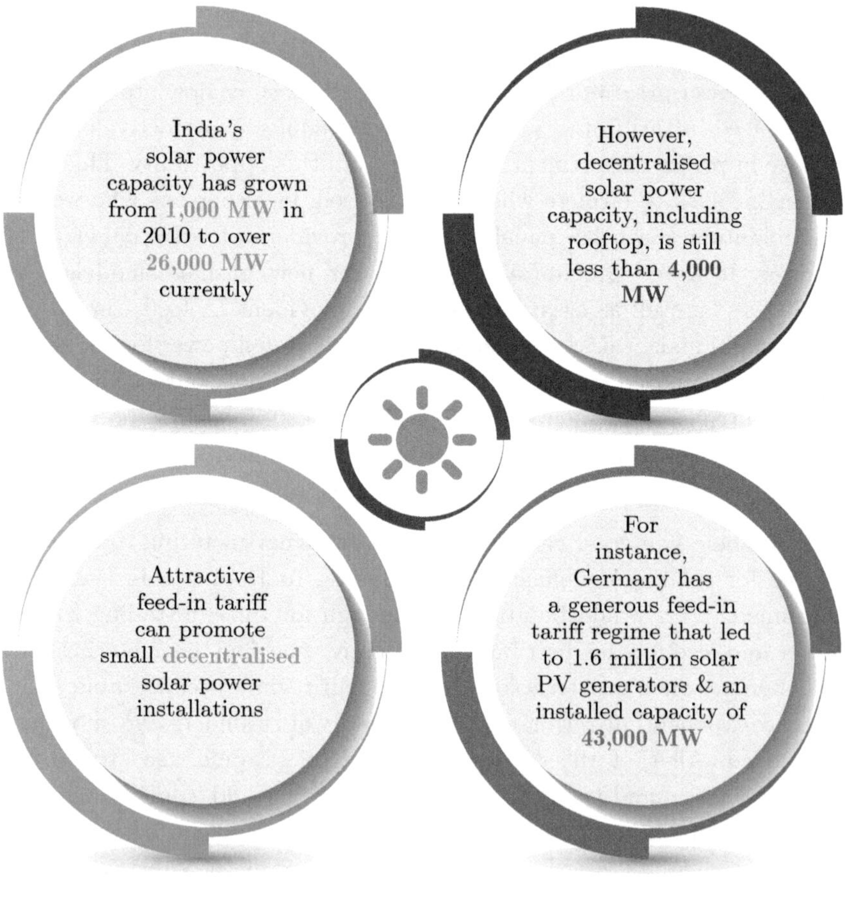

daytime farmers need not leave their pumps on and flood their fields. Other electricity driven technological modernisation in agricultural operations would also become feasible.

About 18% of overall electricity consumption in the country is in agriculture. This could shift to solar power in a few years. Further, the distribution companies would be saving on power purchase costs for meeting the demand for agriculture. As the burden of subsidy for the supply of practically free electricity for agriculture in most states is to be met by the state governments, this would reduce the subsidy burden on their budgets by 25%–30%.

India has 640,000 villages. With just 1 MW of solar power being installed in each village, India could look at having 640,000 MW of solar power. This would be a truly remarkable achievement.

*Mr Ajay Shankar is Distinguished Fellow at The Energy and Resources Institute*

This article appeared in *Business Standard* on December 30, 2018.

India needs a time-bound plan to close old thermal power plants that are contributing to environmental pollution

# Enhancing the Green Footprint of the Electricity Transition in India

*By Rakesh Nath and A K Saxena*

Electricity is a major energy carrier in the world today and its use is expected to increase, while climate change drives increasingly renewable-based electricity to replace fossil fuels in many applications. However, the transition to renewables will occur over time, and there is a clear need to reduce emissions from this sector. The power sector contributed 50% of the 10,500 kilotonnes (kt) of sulphur dioxide ($SO_2$) emissions, 30% of nearly 73,300 oxides of nitrogen ($NO_x$) emissions and around 8% of 6,330 kt of particulate matter (PM) emissions globally. $SO_2$ causes acid rain, $NO_x$ creates atmospheric ozone, and localised emissions of PM from coal-fired power plants lead to respiratory diseases and human mortality. In India, the power sector is responsible for ~53% of $SO_2$ emissions and ~44% carbon dioxide ($CO_2$) emissions.

Consequently, while coal-based power generation is providing energy security to India; it has several environmental and human health impacts, including outdoor air pollution.

## Air Pollution and Health Impacts

Exposure to air pollution is linked to many health effects that vary with pollutant concentration. Air pollution has led to over 2,750 cases of deaths or severe illnesses per lakh people in 2016. It was the second leading risk factor in India's disability-adjusted life years (DALYs) after child and maternal malnutrition, causing 6.4% of India's total DALYs in 2016. Chronic respiratory disease, largely caused by air pollution, is the second largest cause of death after cardio-vascular diseases.

Studies have already looked at the role of coal-based thermal power plants (TPPs) on emissions of PM, sulphur oxide ($SO_x$), and $NO_x$ concentrations, but there is limited bottom-up analysis on dispersion of emissions from the coal-fired power plants. In its Energy Transitions project TERI is using this approach for analysis.

## Potential Impacts on Land and Water

Along the value chain of coal power generation, a large amount of water is consumed and waste water generated. Key environmental impacts include the use of water resources for steam production and cooling, the subsequent discharge of waste-water into water bodies, generation of hazardous waste, and land use for power generation. TPPs account for 87.8% of total industrial water consumption in India. In addition, scarcity and competition with local communities in water use and abstraction can occur.

Coal mining causes environmental pollution, during the extraction and the transportation phases. Frequent coal mining causalities include farmland and ecosystems' diversions, and damage to land resources. Release of Fly-ash residues can contaminate soils and reduce agricultural productivity. All these adverse impacts can be mitigated with transitions to increased renewables in the energy mix.

Laws such as the Right to Fair Compensation and Transparency in Land Acquisition, Rehabilitation and Resettlement Act, 2013, and the Forest Rights Act, 2006, exist to govern the acquisition of land for coal mining or setting up power plants. They are designed to enable adequate compensation to landowners. However, they can cause such delays in land acquisition that the cost overruns become significant for the power generation companies; and land owners are dissatisfied by compensations.

## Greening the Footprint of Thermal Power Stations

At this point, how can India move along its electricity transition and simultaneously reduce the externalities on land, water and human health? Three realities must be considered when attempting to answer this question:

(i) The New Environmental Norms for Thermal Power Stations notified by the Ministry of Environment, Forest and Climate Change (MoEFCC) in 2015 are in place, laying out new emissions standards to control $SO_x$, $NO_x$, mercury and PM emissions, as well as new water norms. Though there are delays in implementation of norms, it is a step in the right direction.

(ii) As per the Central Electricity Authority's assessment, a coal-based capacity of 5,927 MW is being considered for retirement in a phased manner by March 2022, and a broad consensus exists that this can be expected to happen. Further, a coal capacity of 16,789 MW with insufficient capacity to control $SO_x$ emissions has been identified and may have to be retired by 2022 in view of the MoEFCC norms. This would bring the total retired coal capacity up to 22 GW by 2022.

(iii) The coal fleet in India today is under stress. Currently, about 40-75 GW of thermal capacity is stranded or close to stranded, representing an investment value of the order of \$40–\$70 billion. A combination of demand over-estimation, accelerated end-use energy efficiency, imprudent corporate practices, inadequate coal availability from domestic sources, high cost of imported coal, problems with distribution companies (discoms) off-take and payment have led to this situation of excess coal-based generation capacity.

In this context, four concrete suggestions, if implemented, might move us towards a greener footprint in the Indian electricity transition:

(i) ABCD (Advantages, Benefits, Constraints, and Disadvantages) analysis is required to identify generating plants in the country that are well past their useful life, produce little power and/or contribute significantly to environmental pollution. This would enable the drawing up of time-bound action plans for the closing of these old inefficient units in a phased manner, in consultation with the state governments/stakeholders, duly keeping in mind that the electricity requirements in the states do not get impaired.

(ii) At present, in the Fuel Supply Agreements of coal-based thermal power plants, the Annual Contracted Quantity (ACQ) is determined on the basis of installed capacity of power plants and Station Heat Rates (SHR), as specified for the type of generating units by the Appropriate Regulatory Commission. Thus, less efficient plants get more coal leading to higher emissions. It is suggested that the ACQ may be linked to SHRs and distance from coal mines, with a view to improve the overall efficiency in utilisation of coal and minimise emissions.

(iii) A shift in narrative must occur to focus on the environmental benefits emanating from an upgradation to super-critical technology. In order to monitor and quantify this, an innovative indicator could be thought of such as avoided DALYs per MW, which would be computed annually for each plant. The current award scheme recognising the meritorious performance of TPPs could be revised to include avoided DALYs/MW.

(iv) Reducing water footprints of TTPs and achieving water neutrality of their operations has a significant potential to reduce water related stress among different sectors within a region. TPPs, therefore, need to be incentivised to achieve more than the prescribed standards of water consumption, which could be in the form of subsidies on water tariffs, award systems to acknowledge the water-conscious power plants, and incorporation of watershed management activities, specifically within the ambit of Corporate Social Responsibility activities.

## The Way Forward

While a transition to renewables in the long run is much needed, short to medium term strategies need to enable cleaner coal-based electricity generation through realistic assessments of the deployment of pollution abatement technologies; ensure pollution/emission control devices are in place and functional as planned; enable cost recovery through tariffs; as well as create innovative financing mechanisms for high and differentiated capital investment.

Reducing air pollution from the power sector will require installing emission control systems (ECSs) coupled with technological advancements in coal power generation in order to reduce emissions. It will require overcoming several challenges, particularly in old power plants, including initial capital investment, space limitations, restriction in power generation during the installation of ECSs, and selection of appropriate technology. The power plants which are under construction may have to revise their plans to be compliant with new environmental norms.

These initiatives need to be taken by the power producers as well as regulators. Taking note of the requirement of additional capitalisation on account of implementing the revised emission standards in new as well as existing generating stations, the Central Electricity Regulatory Commission has put

The power sector in India is responsible for

- 53% of $SO_2$ emissions and
- 44% of $CO_2$ emissions

- Air pollution was the second leading risk factor in disability-adjusted life years (DALYs) in 2016
- Outdoor air pollution caused 6.4% of India's total DALYs in 2016.

in place a process and a basis for tariff determination in the draft Tariff Regulation, 2019, to be effective from April 1, 2019. This would set at rest the apprehensions that developers have had in this regard. Thus regulatory support to policy is on the anvil, and India's shift to clean power remains unabated. Policy and regulations can support opportunities which create greater benefits from improved performance, than norms do. This could accelerate improved environmental performance.

*Mr Rakesh Nath is former Chairperson of the Central Electricity Authority, and former Technical Member of the Appellate Tribunal for Electricity*
*Mr A K Saxena is Senior Fellow and Director at The Energy and Resources Institute*

# Bibliography

Bhattacharia, S., (2017). *Benefit cost-analysis: Outdoor Air Pollution*, Rajasthan Priorities, Copenhagen Consensus Centre. Retrieved from: https://www.copenhagenconsensus.com/sites/default/files/raj_outdoor_air_pollution_formatted.pdf

Central Electricity Authority, MoP, GoI, (2018). *National Electricity Plan.* Retrieved from: http://www.cea.nic.in/reports/committee/nep/nep_jan_2018.pdf

Enerdata, (2017). *Global Energy and $CO_2$ Database.* Retrieved from: https://www.enerdata.net/research/energy-market-data-co2-emissions-database.html

India State-level Disease Burden Initiative Collaborators, (2017). Nations within a nation: variations in epidemiological transition across the states of India, 1990–2016 in the Global Burden of Disease Study. In *The Lancet, Vol.390, Issue 10111.* Retrieved from: https://www.sciencedirect.com/science/article/pii/S0140673617328040?via%3Dihub

International Energy Agency, (2016). *Energy and Air Pollution: World Energy Outlook Special Report.* Retrieved from: https://www.iea.org/publications/freepublications/publication/WorldEnergyOutlookSpecialReport2016EnergyandAirPollution.pdf

Grover, S. and Tayal, S., (2016). *Water Neutral Electricity Production in India: Avoiding the Unmanageable.* Retrieved from: http://www.teriin.org/policy-brief/water-neutral-electricity-production-india-avoiding-unmanageable

Guttikunda S K, Jawahar P., (2014). Atmospheric Emissions and Pollution from the Coal-fired Thermal Power Plants in India. In *Atmospheric Environment, Vol. 92.* Retrieved from: https://www.sciencedirect.com/science/article/pii/S135223101400329X

Roshna, N. and Srinivasan, S., (2018). *Financial Implications of Emission Standards for Coal Power Plants.* Retrieved from: http://www.cstep.in/uploads/default/files/publications/stuff/CSTEP_BCA_Policy_note.pdf

Hydrocarbon molecules derived from solid biomass, liquid biofuels or biogas could provide a rapid route to decarbonise "hard-to-abate" sectors

# Alternative Fuels for a Zero-Carbon Economy

*By Adair Turner*

Huge falls in the cost of wind and solar electricity – down 60% to 90% over the last 10 years – have dramatically changed the economics of power generation. Work by the Energy Transitions Commission shows that renewables will increasingly be a cheaper power source than coal, and that power systems could run efficiently even if 80% to 90% of electricity came from intermittent renewable sources. In a developing country like India, huge increases in electricity demand – up perhaps 2½ times within the next 15 years – can be met without further coal power investment. This will deliver huge local environmental benefits, cutting the air pollution which threatens health in many Indian cities, and would help limit the global warming to which India is more vulnerable than almost any other country. Decarbonising power generation and electrifying as much as the economy as possible is the core of any path to a zero-carbon economy.

## Challenges to Electrification of Energy

While many uses of energy are already electrified (such as air-conditioning systems and factory machinery) or can easily become so (with electric cars bound to dominate future road transport), it is less easy or more expensive to electrify some heavy industrial sectors (such steel or cement) and currently impossible to electrify international aviation. If the world is to achieve the zero-carbon economy we need to prevent harmful global warming, we must find a way to take the carbon out of these 'hard-to-abate' sectors.

The fundamental challenge is two-fold. First that it is much easier to turn fossil fuel molecules into the intense heat which many industrial processes require than to generate that heat with electricity. Second that hydrocarbon molecules have far higher energy density than batteries: about 11 kilowatt hour (kWh) of energy is stored chemically within a kilogram of diesel fuel, versus only about 0.3 kWh in a kilogram of lithium-ion battery. Unless and until that battery density increases at least 6 times, battery flight from Delhi to London will continue to be impossible.

Despite these challenges, the role of direct electrification is likely to grow even in some of the apparently more difficult sectors. Recent work by the Energy Transitions Commission shows that battery-powered trucks will become increasingly feasible and cost competitive during the 2020s and electrifying short and medium distance trucks could deliver hugely beneficial air quality improvements in India's major cities. In principle cement kilns can be electrified and the cost of doing so will reduce over time: and recycled steel is already produced using electric arc furnaces rather blast furnaces using coking coal. But to achieve full decarbonisation of these hard-to-abate sectors will require new low-carbon fuels.

## Decarbonising 'Hard-To-Abate' Sectors

There are two main alternatives. One is to derive hydrocarbon molecules from biomass sources, using them for combustion purposes either as solid biomass (e.g. in cement kilns), liquid biofuels (in trucks and planes) or as biogas (for instance to meet peak electricity demand). The other is to produce zero-carbon hydrogen – whether using zero-carbon power to electrolyse water or from steam methane reforming combined with carbon capture – which can then be used in fuel cell vehicles to drive electric motors, combusted to produce heat and drive turbines, or used as an alternative to coking coal as a reduction agent in steel production. Multiple hybrid solutions are also increasingly possible: for instance, hydrogen injection could significantly increase the amount of biogas produced from any given quantity of biomass.

Together with direct electrification, these options make it possible to decarbonise all sectors of the Indian economy at acceptable costs. Direct electrification will likely dominate urban buses and short-distance trucking, but hydrogen fuel cell vehicles may play a major role in the long-distance truck sector. Electrifying India's huge railway system should be a high priority, but for some routes using hydrogen as an energy source may be more economic.

Burning municipal waste or agricultural residues could provide a rapid route to decarbonise cement production, and India's Dalmia Cement, already one of the lowest carbon cement producers in the world, is committed to being carbon negative by 2040. And with much of Indian steel production already using a syngas-based 'direct reduction' technology, switching to hydrogen-based reduction would be an easier transition than in countries where coal-based blast furnaces dominate.

In aviation meanwhile, both batteries and hydrogen are likely to play an increasing role in short distance flight, but flights which I take from London to Delhi will almost certainly need to be powered by a precise bio or synthetic equivalent of existing jet fuel. This is undoubtedly technically possible even if higher production costs will mean a moderately higher ticket price.

## How To Decarbonise?

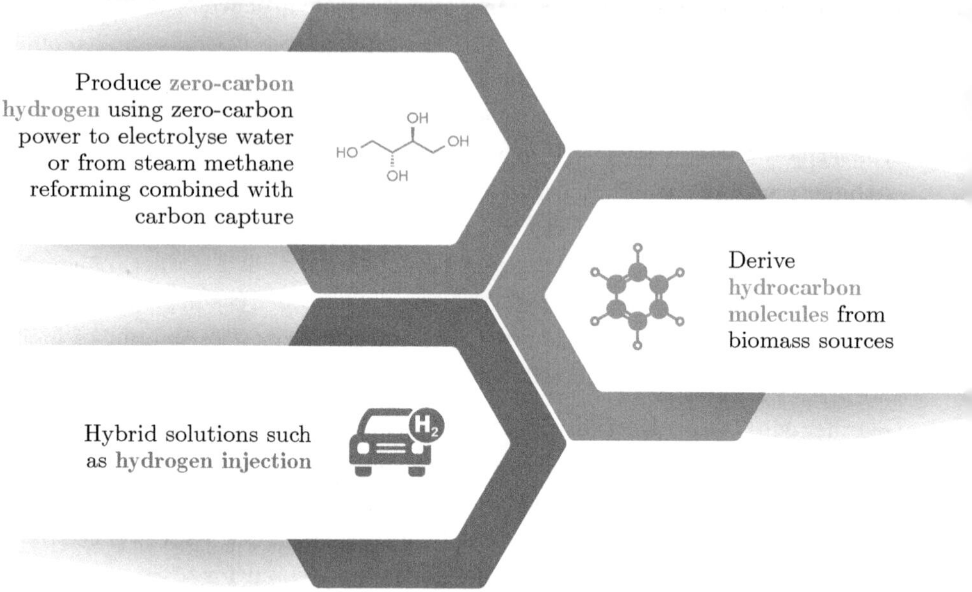


## The 'Sky Scenario'

In India as across the world, the most cost-effective route to a zero-carbon economy is thus certain to include 'green molecules' as well as 'green electrons'. At the global level, 'green electrons' will play the dominant role. As the 'Sky Scenario' which Shell has recently published makes clear, any credible route to a zero-carbon economy will see direct electricity use rising from around 20% to 25% of final energy demand today to something like 60% by 2070. Some continued fossil fuel use – perhaps 10% to 15% of final energy demand, might also still be possible if Carbon Capture and Storage or Use can be made a cost-effective and safe technology.

But in India as elsewhere, a hugely expanded role for hydrogen will be vital, and hydrocarbons produced from biomass sources will also be an essential part of a cost competitive and environmentally sustainable energy mix.

*Lord Adair Turner is Chair of the Energy Transitions Commission*

# MAKING THE GREEN SHIFT RELIABLE

Given the daily and seasonal variations of renewable sources, India needs to enhance power system flexibility to balance demand and supply on a real-time basis

# Electricity System Flexibility in the Context of the Transition to a High Share of Renewables in India

*By Thomas Spencer*

India faces the significant challenge of providing access to affordable, secure, and low-carbon energy to its people. Within the broader energy sector, electricity is crucial to the information-driven modern economy. No country has grown to the levels of income to which India aspires without providing secure and affordable electricity to its people. Now, with the advent of climate change and other environmental stresses, we must add "sustainable" to the objectives of the electricity sector.

It is noteworthy that India is gaining global recognition thanks to its voluntary commitment to achieve an energy mix that promotes low-carbon growth. Characterised as one of the world's largest renewable capacity expansion programmes, India targets 175 GW of renewable energy (RE) capacity by 2022. This includes 100 GW solar, 60 GW wind and 15 GW of other renewables, including small hydropower. Additionally, India has pledged to increase the share of non-fossil fuel generated electric capacity to 40% by 2030.

This massive commitment to energy sustainability comes with its own set of challenges. For instance, the seasonal and daily variability of RE sources creates the necessity for an increasingly flexible electricity system. The stability of the electricity grid requires supply and demand to balance each other in real time, without which blackouts, brownouts, equipment damage, disrupted economic activity and social welfare occur.

## Grid Flexibility and How to Achieve it

This is what flexibility is about. Grid flexibility – which can be achieved through a portfolio of options including storage, Demand Reduction Measures, and utilisation of existing thermal capacity – refers to the capacity of the grid to balance demand and supply in real time, while dealing with planned and unplanned variations in either demand or supply. While the requirement for flexibility is not just a function of renewables, they do increase the need for it because of their non-dispatchable and variable nature. In addition, a number of India-specific challenges must be overcome for a smooth transition to the flexible electricity system required to integrate a high share of renewables:

1. **High reliance on solar as a source of renewable power:** Solar energy is a huge and cost-effective resource in India. The country has 250-300 sunny days in a year, and solar tariffs have fallen to less than ₹3/kilowatt hour (kWh), making it competitive with coal. However, solar power can only produce electricity during the daytime, and requires significant backup capacities to operate at night. Storing the solar energy of the day and releasing it to the grid at night is a solution, though it cannot be done economically (yet). It

is estimated that by 2022, the grid will need to provide around 25 GW of 'fast-ramping' capacity (equivalent to about 38 large coal units) during the night, as a back-up to solar when the sun sets. While the power grid has this capacity, its utilisation will require more flexible operations of coal, gas and hydro plants to follow the cycle of demand and solar power production. In the longer-term, large scale battery storage will be essential to significantly increase the share of solar in the grid. Fortunately, the cost of storage is dropping rapidly, and it should be economically viable in the mid-2020s.

2. **Seasonal variability of wind and hydropower generation:** Power generation from wind and hydropower in India depends heavily on the monsoon season, which typically lasts from July to September, when the wind output is much higher. This leads to the need for seasonal flexibility during winter, when generation from wind and hydropower is low. A challenge is that those backup capacities available during the winter months would likely be lying idle during the monsoon months. As low utilisation of capital raises costs, developing low-cost options for seasonal flexibility will be crucial.
3. **The growing importance of the air-conditioning load as a driver of evening and night-time peak demand:** The 'non-coincidence' of this night load with the daily solar output will raise many challenges for the Indian grid, unless storage systems can be deployed at scale. Currently, all-India electricity demand peaks at night, particularly in summer when households switch on their air conditioners. However, air-conditioning penetration in Indian households is still below 10% and is expected to increase massively as incomes rise, adding to the night peak challenge. Solar power coupled with storage technologies can provide solutions.
4. **Immaturity of electricity markets:** Electricity markets account for a very small share of generation and are still at a nascent stage in India. Experience from high-renewable countries or regions like Germany and California shows that integrating a high share of renewables is greatly facilitated by improving markets three-ways:

a. Markets need to be 'faster' (closer to real time, rather than hours or days ahead of electricity delivery). This enables the better coordination of production from other sources to compensate for the variability of renewables, which is more predictable closer to real time.

b. Markets need to be 'bigger'. The correlation between different renewables resources increases with increasing geographical areas (the wind is always blowing somewhere). Thus, markets which are able to balance electricity demand and supply across larger geographical areas are better able to deal with the variability of renewables. In India, this means a progressive shift away from mostly state-level markets, towards regional and national level ones, and a simultaneous strengthening of the electricity transmission system. This will require inter-state cooperation in grid planning and operations.

India targets 175 GW of RE capacity by 2022. This includes 100 GW solar, 60 GW wind and 15 GW of other renewables, including small hydropower

The country has 250-300 sunny days in a year, and solar tariffs have fallen to less than ₹3/kWh, making it competitive with coal

It is estimated that by 2022, the grid will need to provide around 25 GW of "fast ramping" capacity (~38 large coal units) during the night, as a back-up to solar when the sun sets

Integrating a high share of renewables is greatly facilitated by improving markets to be:

- 'Faster'
- 'Bigger'.
- To give better 'price signals for flexibility'

c. Markets need to give better 'price signals for flexibility'. In a high-renewables system, flexibility is the core. Resources that are providing this flexibility must be adequately incentivised. The market reform agenda in India is vast, but the government is aware of this and is initiating several reforms.

5. **Need for significant investment:** Ramped up grid flexibility requires substantial investment, for example for retrofitting existing power plants, developing pumped hydro and battery storage capacities, and strengthening the transmission infrastructure. At the same time, it is important to ensure that plants remain economically viable and power prices do not burden consumers. After all, affordable power is a crucial aspect of India's 24X7 'Power for All' vision. Studies on the 'grid integration costs' of renewables are scarce in India, and much needs to be done in this regard.

## Recent Reforms and Regulations

Despite these challenges, India is making progress towards enhancing the grid's ability to adapt to changing electricity supply and demand patterns. To enhance grid flexibility, the Government of India is progressively implementing new power sector reforms and regulations. In 2016, the guidelines for flexible utilisation of domestic coal were released. Those aimed to avoid unnecessary transportation of coal to reduce the cost of power generation. Recently, the Ministry of Power decided that there should be flexibility in generation of power and scheduling of thermal power stations, so that distribution companies can meet their Renewable Purchase Obligations without any additional financial burden. This flexibility will provide power generators the opportunity to optimally utilise generation from renewable sources, and also contribute to carbon emissions reduction. The Central Electricity Regulatory Commission has initiated activities on electricity market reform, including on developing a real time balancing market, a market for ancillary services, and shifting from 15 to 5 minute operational time blocks for the scheduling and dispatching of electricity. All of these steps are in the right direction towards the profound changes required in India's grid to integrate renewables.

As India marches on the path to a flexible energy system, it becomes crucial for the country to develop an integrated approach where future-oriented solutions such as cross-border energy exchanges, energy storage, demand-side measures and effective markets are at the core. This situation, too, presents us with an opportunity to unite efforts from all quarters and transform the energy scenario of the country.

*Mr Thomas Spencer is Fellow at The Energy and Resources Institute*

---

## Bibliography

Mangotra. K., Agarwal, S., Hillbrand, A., Kwatra, S., Deol, B., Awasthi, A., Jaiswal, A., Andersen, S., O., Sherman, N., J., & Zaelke, D., (2018). *Improving Air Conditioners in India.* Retrieved from: http://www.teriin.org/sites/default/files/2018-04/improving-air-conditioners-in-india.pdf

Ministry of New and Renewable Energy, GoI, (2017). *Annual Report 2016-1017.* Retrieved from: https://mnre.gov.in/file-manager/annual-report/2016-2017/EN/pdf/1.pdf

Ministry of Power, GoI (2018). *Flexibility in Generation and Scheduling of Thermal Power Stations to Reduce Emissions.* Retrieved from: https://powermin.nic.in/sites/default/files/webform/notices/Flexibility_in_Generation_and_Scheduling_of_Thermal_Power_Stations_to_reduce_emissions.pdf

High variability of demand and supply are a challenge for power markets. Energy storage, power pools and energy saving certificates will help address these challenges

# Accelerating Reforms in Power Markets in India: Key Trends, Challenges, Recommendations and Opportunities

*By Hemant Sahai and Rachika A Sahay*

## Introduction and Key Trends

The electricity market establishes a price for its commodity through the basic economics principle of demand and supply. However, an electricity market produces the commodity for immediate consumption even though the requisites for creation and classification of the markets such as quality, quantity, price etc. remain the same as those of other traditional markets. In the Indian scenario, the power exchange has matured rapidly in spite of initial low volumes and high prices. Though bulk of the electricity continues to be contracted through bi-lateral long-term power purchase agreements (PPAs), the 'electricity market' refers to trading at the power exchange.

The markets within a power exchange are spot markets, where the trading takes place in real time, and forward/derivative markets, where the delivery of the commodity takes place at a future date. For electricity, all factors influencing its supply and demand have an immediate impact on the price in the spot market.

## Challenges and Related Recommendations

Challenges arise because both supply and demand are highly variable. Recently, renewable energy integration into the grid has been unprecedented. The variability of renewable sources introduces uncertainty in generation output.

Demand is equally unpredictable in the market, stemming from the fact that end consumers have started acquiring distributed energy sources making them self-sufficient in the production of their own electricity, through the use of rooftop solar panels for example. This reduces demand for electricity in the market. Conversely, certain novel uses of electricity such as electric vehicles and battery charging can also increase the demand.

These are constraints which the electricity market has to work with. However, Demand-Side Management (DSM) measures can be an effective stepping stone to counter these challenges. This involves planning, implementing and monitoring of various activities that are designed to modify consumers' electricity consumption patterns, both with respect to timing and amount of electricity consumed, through encouraging optimal and efficient use of electricity. This includes (i) energy storage; (ii) power pools and (iii) energy saving certificates (ESCert).

(i) Investing in the research and development of both battery-based and utility scale storage systems, including pumped hydropower, flywheels and supercapacitors, lithium-ion batteries, etc., would help the Indian power market deal with many challenges head on. The paper issued by the Central Electricity Regulatory Commission (CERC) discussing tariff determination models for multiple users of Battery Energy Storage Systems (BESS), commercial viability of BESS, and policy changes required to deploy bulk storage facilities in the country is encouraging to understand the holistic process of tariff-setting. Energy storage is a critical distribution asset that addresses challenges in peak load management and frequency regulation.

(ii) Developing competitive power pools (Pools) can respond to challenges posed by spot markets. Pools are a system where the output from different power plants is pooled together in order to coordinate operations and planning and meet demand more reliably. A bidding system could determine the purchase and sale price of electricity which would respond to imbalances between scheduled and actual supply. Pools could largely function as self-regulatory bodies if they are kept independent of other market participants – avoiding conflicts of interest – if appropriate frameworks for operating principles are formulated, if requirements by its participants are met, and if efficient rules for determining Pool prices are established. Further, de-mutualisation would establish credible corporate governance of the Pools. Pools would not be subject to regulatory oversight of CERC as they do not trade electricity. Instead, they would fall under the jurisdiction of Securities and Exchange Board of India under the Securities Contracts Regulation Act, 1956. Such a Pool structure would ensure that power producers, decision makers and consumers are kept distinct groups; that the shareholders are responsible for key decisions via an independent committee; and day-to-day decisions are made by professional managers to keep the sanctity of the exchanges intact.

(iii) ESCerts are awarded to energy intensive units that overachieve their energy reduction targets under the Perform, Achieve and Trade (PAT)

Scheme of the Government of India. Each certificate is equal to one metric tonne of oil. Units that could not meet their assigned targets were required to purchase ESCerts from the overachievers. Over PAT Cycle I, a total of 38,50,000 ESCerts were awarded and 110 facilities were directed to purchase 14,50,000 ESCerts for non-achievement. But when the trading began, the opening price of each ESCert was ₹1,200 instead of the expected price of ₹10,000. Consequently, achieving the targets turned out to be more expensive than compensating for non-compliance. To ensure that ESCerts promote energy efficiency, the regulator must set a minimum price that encourages energy intensive industries to invest in energy efficiency, rather than buy ESCerts.

Transmission congestion is another challenge faced by power markets. This occurs when the power output is higher than the transmission system's capacity. While the government has commissioned 13,820 circuit kilometers of transmission lines during 2017-18, this is 59.9% of the annual target fixed for that period. Furthermore, in a majority of Indian states, distribution licensees dominate the power distribution network. Considered the weakest players of India's power sector, the distribution companies (Discoms) face many challenges including controlling aggregate technical and commercial losses, ensuring financial viability, providing electricity access to all households and reducing inefficiencies in power generation and planning. While the Discoms are under severe stress, the vital need to develop the distribution network is often neglected.

The efficient implementation of improved transmission capacity as proposed by the government will relieve the overall congestion instead of relieving a specific constraint until another one becomes constraining. Indian power markets may adopt zonal/regional trade or integrated market models until the transmission capacity gets augmented across the country.

The Electricity Amendment Bill of 2014 seeks to segregate carriage and content of electricity by transferring the supply function from the Discoms to the suppliers. This separation encourages healthy competition and allows the consumers to buy electricity from the supply they choose. In addition, this would enable the over-burdened and financially unstable Discoms to focus on strengthening the distribution network rather than on procurement and supply of electricity. The government should strive for the successful implementation of separation of carriage and content, as it will relieve the power sector.

## Opportunities

While there are many opportunities in power markets, this article focuses on a few of them. Prior to deregulation of the sector, the electricity market was vertically integrated, and the price of power was regulated by state-owned authorities. Consequently, post-deregulation prices began to be determined by demand and supply, which as we have seen, are variable. This has translated into price volatility as the market becomes more competitive. Those fluctuations can be hedged with derivatives which derive value from underlying assets. For electricity the underlying asset is its price. Future reforms of trading instruments

- Variability of renewable sources causes uncertainty in output
- Unpredictable demand in the market
- Transmission congestion
- Discoms' financial constraints

- Demand-Side Management (DSM) measures:
  (i) energy storage
  (ii) power pools
  (iii) energy saving certificates (ESCert)
- Improved transmission capacity
- Separation of carriage and content

in Indian power markets should give an important role to derivatives in providing price discovery and facilitating effective risk management. However, to implement these instruments, certain structural changes to the sector need to be made. This includes better communication infrastructure and augmenting transmission capacities.

India became a net exporter of electricity for the first time in 2017, exporting around 5,798 million units to Nepal, Bangladesh and Myanmar. With the introduction of 'Guidelines on Cross-Border Trade of Electricity' in December 2017, India took a first step towards formalizing this cross-border market including at the power exchange. However, many aspects of the Guidelines remain to be addressed. Hopefully detailed regulations on cross-border trade of electricity are issued soon, as market participants are counting on CERC to establish conditions for participating in such a market, including through the power exchange.

*Mr Hemant Sahai is Founding Partner at HSA Advocates*
*Ms Rachika A Sahay is Partner at HSA Advocates*

As more consumers also become producers of electricity, one-way flows will need to evolve into multi-directional flows of electricity and digital information

# **Decarbonise, Decentralise, Digitalise** – the Future of Energy in India

*By Sunil Mathur*

India is among the world's high-growth economies, and as its share in the global gross domestic product (GDP) continues to rise, so does the demand for reliable and uninterrupted energy supply. Besides being among the world's top five producers of electricity, hydro-electric power, coal and wind energy, India is also among the top consumers of energy.

While its population and sustained economic growth is driving energy consumption, India is also faced with challenges such as an aging energy infrastructure (with some 35 GW-40 GW of power plants nearing their end of life), uncertain oil prices that make overall energy costs unpredictable, blackouts due to natural disasters or other events, which raise concerns over grid resilience and energy security of systems. Besides, developments in power generation, distribution, storage and consumption are making the energy landscape more and more complex.

These challenges redefine the way we look at energy systems and this is where three key disruptive elements – decarbonisation, decentralisation and digitalisation – come into play.

## Decarbonisation

Governments and corporates around the world are under tremendous pressure to reduce their carbon footprint and ultimately turn carbon neutral. Utilities are perennially challenged to enhance production efficiency and bring down the cost of electricity. This is driving the growth of renewables into the mainstream energy mix to make it green and sustainable. As per the Ministry of Power, renewable energy sources in India contribute around 20% to the installed power capacity, second after coal. Coal-based power generation in India caters to approximately 70% of the energy supplied to the grid, contributing around 2,000 million tonnes of $CO_2$ to the environment. However, over time the carbon intensity of the Indian power sector is expected to decline significantly, as the share of renewables increases significantly and older, less efficient coal plants are retired. The energy intensity of the economy has fallen by 13% during the period 2012-2017 as against the G20 average decline of 11% during the same period. The target of 175 GW of renewables by 2022 implies that the share of variable renewables in the power mix increases from the current level of about 7% to about 20%. The variable nature of renewables is forcing conventional energy sources to continue at partial load and become more flexible in their operations to serve the base load requirements of the grid.

Technology can enable fast ramp-up or ramp-down of thermal units, part load operation, efficiency and reduced carbon emissions. Power-to-gas solutions produce green hydrogen that has applications across energy storage, re-electrification and mobility. It will be a significant contributor to facilitate cross-sector coupling. For corporates and governments committed to carbon neutral targets, these are effective solutions. Changes in energy systems will also lead to the requirement of storing energy for extended periods of time, especially when the reliance on renewable power sources increases. One example of suitable infrastructure would be gas grids with their tremendous storage potential. Other emerging trends mean that the location of energy generation will shift. Rooftop solar PV, electric vehicles, and battery storage are emerging as technologies that will significantly disrupt the distribution grid in the coming 10-15 years.

## Decentralisation

That brings us to the second 'D' – decentralisation. As more and more sources of electricity generation and storage get added to the distribution grid, consumers will become 'prosumers' (i.e. producers and consumers). One-way flows will evolve into multi-directional flows of electricity and digital information. This adds complexity to the grid and increases reliability issues. It will be almost impossible to intervene on and correct these issues manually. Therefore, the networks need to be analytical, intelligent and self-healing to sustain the balance dynamically in real time. India, therefore, needs to invest in making the grid resilient and intelligent.

There will be a need to provide localised solutions to emerging challenges like renewable energy integration, electric vehicle charging infrastructure management, network stability, power reliability, and load and demand management. Additionally, digital sensing and automated analytics-based solutions will enable the move towards a more efficient, reliable, resilient and responsive grid.

## Digitalisation

With a rise in the policies for energy trading and rapid development of energy

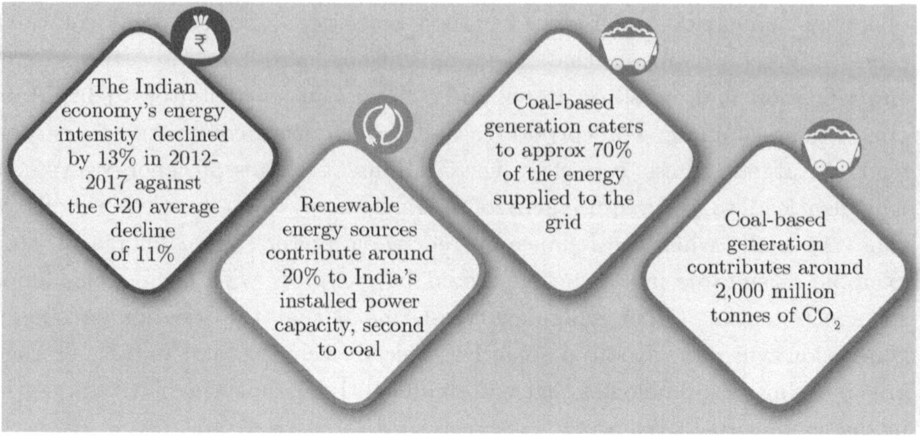


storage solutions, more and more prosumers will use the grid for trading. This will give consumers the option to choose the electricity source based on their objectives. For example, consumers who prioritise costs will be able to choose the cheapest electricity source and those who prioritise sustainability will be able to choose from a renewable/green source which will be enabled by evolving technologies such as blockchain.

These developments will make the grid extremely complex to manage manually and therefore digitalisation will be the key. The exchange of energy will witness the exchange of data. This big data and price analytics will form the basis for net metering and meter data management in a cyber secure environment. Self-healing network solutions in the distribution network, by ensuring automatic fault localisation, isolation and restoration will allow re-supply of power in less than a minute in the event of an outage.

Digitally connected equipment will open up possibilities like remote monitoring and operations. Turning big data into a digital twin by leveraging smart data, analytics and machine learning will enable the efficient running of equipment and improved life cycle management while paving the way for predictive maintenance. For example, e-mobility infrastructure management solutions will allow charging station management, execution of customer contracts and services. It will also enable management of electric vehicle charging in a smart way based on real-time demand on the grid coupled with the price of energy.

Digitalisation will also drive cost-efficiency with digital twin solutions leveraging automation of end-to-end processes in power plants (from coal yard management to ash utilisation) or an entire off-shore platform being managed remotely, efficiently and with less people.

Industrial Internet of Things (IoT) has already arrived and is accepted as a platform to build applications. The stage is now set for widening the scale of digitalisation to fundamentally alter business models such as the transition from capex to availability or use/benefit-based fees in both operations and maintenance (for example third-party contracting in rooftop solar PV). These changes will usher significant user efficiency and price advantages. The future of our world does seem to be all electric. It will be decarbonised, decentralised and above all digitalised from end-to-end. To be well prepared, the nation should seize the opportunity today.

*Mr Sunil Mathur is Managing Director and Chief Executive Officer of Siemens Limited*

## Bibliography

Climate Transparency (2018). *Brown to Green: The G20 Transition to a Low-Carbon Economy.* Retrieved from: https://www.climate-transparency.org/wp-content/uploads/2019/01/2018-BROWN-TO-GREEN-REPORT-FINAL.pdf

Capital intensive renewable energy projects will require more than $100+ billion investment by 2022 to meet India's capacity targets

# **Financing Clean Energy:** Enhancing Capital Availability and Creating Supportive Frameworks for Renewables in India's Energy Mix

*By Manoj Kohli*

Thanks to a combination of factors like a global concern for energy security, access to clean energy, financial viability, technology advancements, falling costs and supportive policy frameworks, the growth in renewable energy deployment is healthy and increasing year-on-year.

Since 2015, developing economies have invested more in the renewable sector than their developed counterparts. In the FY2017, the global renewable energy industry saw new investments worth over $280 billion, which resulted in over 150 GW of new capacity additions. This was more than the net capacity added through conventional sources where solar and wind saw the lowest ever tariffs quoted under the auction regime in India and abroad.

## Steady Progress on Aggressive Targets

The Government of India has set the target to achieve 175 GW of installed renewable energy by 2022 and, under the Paris Climate Agreement, has committed to have 40% electricity generation from non-fossil fuel resources by 2030. To achieve these, the government has been working aggressively towards scaling up renewable energy capacity, which today constitutes ~20.5% of the generation capacity, mainly led by wind (34 GW) and solar (23 GW) and followed by biomass and small hydro. With the right policy support and regular renewable energy auctions conducted during FY17-18, the power sector saw an addition of 13 GW of renewable capacity. As per the Ministry of New and Renewable Energy (MNRE), the renewable sector attracted an investment of over $42 billion in the last 4 years; but more investments are required to achieve the desired targets.

Capital intensive renewable energy projects will require more than $100+ billion investment by 2022 to meet the capacity targets. With low amounts of public funding, the onus of developing renewable energy projects lies with the private sector. Institutional investors like pension funds, sovereign wealth funds and insurance companies looking for long-term, stable returns can play a critical role in scaling up renewable capacity. However, this will totally depend on enabling frameworks created by the government to attract new investments within the sector.

The Indian government has made efforts to drive renewable energy financing by providing avenues of alternative funding, by creating new institutions, mechanisms and instruments:

- Recognising the renewable energy sector as a priority sector for lending;
- The National Clean Energy and Environment Fund (NCEEF) to help

sectoral entrepreneurial ventures, and research and development (R&D) through a cess on coal;

- Extension of soft loans by non-banking financial companies (NBFCs).

However, the overall development of renewable energy financing remains a challenge and to overcome this, the government needs to focus on the following aspects:

a) **Perceived Risks for the Sector:**

Risk perception of financial institutions and investors creates many uncertainties for investment, which need to be addressed urgently.

- Policy risks: Uncertainties like non-compliance with Renewable Purchase Obligations, sanctity of bidding processes, and non-existence of single-window clearance mechanisms are some of the policy risks.
- Off-Taker risks: Poor financial conditions of the distribution companies (Discoms) resulting in delays in payments, as well as back-down of renewable power in spite of its must-run status, are Off-taker concerns.
- Uncertainty on duties to be implemented on projects; delays in resolution of issues pertaining to changes in clauses of Power Purchase Agreements (PPAs); concerns over grid infrastructure; and high costs of private land etc., are a few other major risks.

b) **The Cost of Funding:**

The cost of funding is critical because a far greater part of electricity from renewable energy is cost-based capital. Due to high interest rates revolving around 10%, and the shorter duration of loan terms of 15-18 years, the cost of debt funding increases substantially. For sustaining the tariffs at the levels discovered during recent renewable auctions, the cost of debt needs to be brought down substantially. A long-term financial market supported by institutional investors could be ideal for renewables, enabling lower cost of debt funding.

Currency risks: With liabilities in foreign currency and operating cashflows in rupees, unexpected changes in currencies can impact returns of projects dearly. We require policies that protect the investors and lenders against currency volatility.

**c) Utilising the Full Potential of the Green Bond Market:**

The promotion of fixed income securities to finance clean energy investments at a lower cost of capital is an attractive investment vehicle for large institutional investors. These are secured against underlying operating assets. Current mechanisms allow green bonds to be revenue-backed, where debt payments are serviced from the cashflow generated by the project. Any entity which can issue a bond, based on the availability of green assets, can issue these bonds; thus, allowing diversified issuers in the market. In India there is low utilisation of green bonds, due to the low trust of investors towards

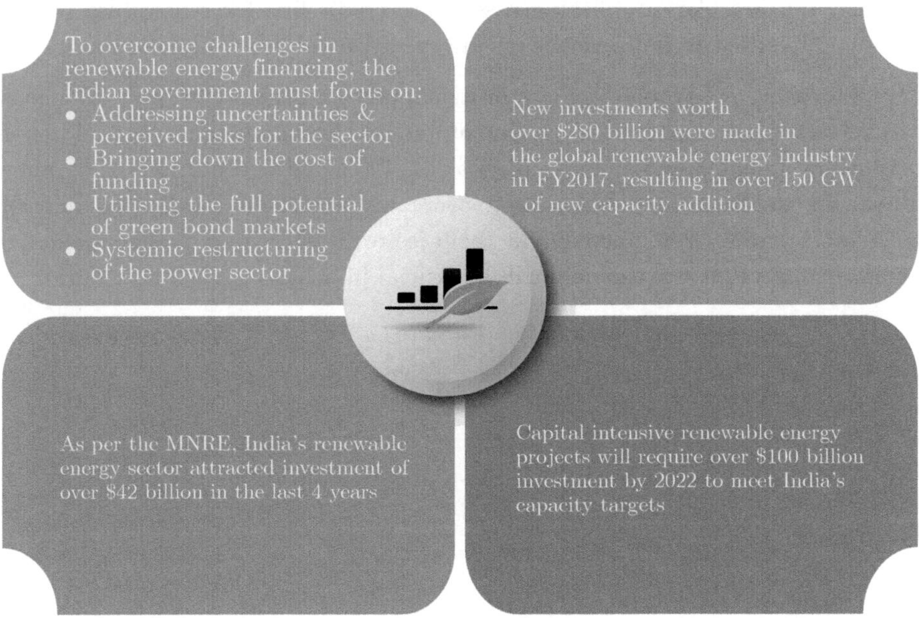

the actual utilisation of the funds raised. There is a need to spread awareness among investors, making the process standardised, and devising incentives for issuers and investors.

d) **Systemic Restructuring of the Power Sector:**

Overall structural issues of the power sector, like poor balance sheets of the Discoms and Independent Power Producers, conventional projects converted to Non-performing Assets, overexposure of banks in the power sector etc. result in a reluctance shown by financial institutions to extend credit lines to new renewable energy projects. The Indian government is striving hard to address some of the key issues like providing facility for uptake of renewable energy through the implementation of green corridors, improving financial health of the Discoms through schemes like UDAY, and forming committee(s) to improve the health of the thermal sector. However, additional policy pushes like establishing a separate credit line for renewables will be required to improve the overall creditworthiness of the sector.

By increasing renewable capacity in its supply mix, India can meet its energy needs to support economic development and growth in a manner that addresses its environmental and climate objectives. To achieve conditions that allow investors and financial institutions to overcome investment hurdles and develop the sector, new instruments and facilities will be required. The proactive engagement of all stakeholders will be crucial in achieving such a scenario.

*Mr Manoj Kohli is Executive Chairman of SB Energy – SoftBank Group*

# MAKING THE GREEN SHIFT VIABLE

India aims to provide free cooking gas connections to 80 million poor households by 2020

# **Bridging the Last Mile:** Accelerating Access to Cooking Energy in India

*By Amit Kumar*

'Energy transition' is a term that has become an intrinsic part of mainstream discussions around energy. However, the transition of energy in such discourses is primarily about transition in the way electricity is, and would be, produced. Let us now imagine what a transition in cooking energy would look like, especially in the rural setting. At a global level, International Energy Agency's Energy Access Outlook 2017 estimates that by 2030, 2.3 billion people will still lack access to clean cooking facilities. As per the Census of India 2011, out of over 240 million households, 100 million were relying on traditional solid biomass, such as firewood and dung cakes, etc., as primary fuels. Fortunately in India, the first phase of this transition has already begun through 'Pradhan

Mantri Ujjwala Yojana' (PMUY) by actually providing LPG (Liquefied Petroleum Gas) connections to 50 million below poverty line households. This target has increased to 80 million households in the Union Budget of 2018.

While the provisioning of millions of LPG connections in such a short span of time has to be lauded, especially as a necessary first step, do connections automatically translate into shifts of households to LPG as a primary cooking fuel? This is a hypothesis worth examining, especially from the point of view of the intended impact in terms of safeguarding the health of women and children by providing them with clean cooking fuel. Recently, it was also reported that the Uttar Pradesh government was working with the NITI Aayog to distribute 20,000 methanol cookstoves to the poor in areas where the Ujjwala scheme had not yet made inroads. Many studies, including The Energy and Resources Institute's (TERI) own primary surveys, have established that mere ownership of clean cooking devices does not automatically mean that the household has moved up the energy ladder. The reasons range from the fuel being free to cooking practices. It underscores the fact that for transition to take place, sustained campaigning around health impacts and other behavioural aspects is equally important.

## Accessibility to Reliable Fuel

While acceptability and affordability of alternative cooking solutions are often discussed as key determinants in their wide scale adoption, accessibility of the fuel too plays a major part in making fuel choices. The fact is that many a time in rural areas, easy availability of solid biomass makes it a fuel of first choice, despite the associated ills. On the other hand, last mile delivery of services or connectivity has been the weakest link in most of the programmes even in metropolitan cities. This, then, assumes a much bigger dimension for villages that are far-off or have scattered populace. A combination of factors, such as geography, poor road networks, and economically weaker consumers render the cost of delivery and traditional distribution model unviable. However, for a cooking solution to be the first choice, reliability of timely supply of fuel has to be absolute.

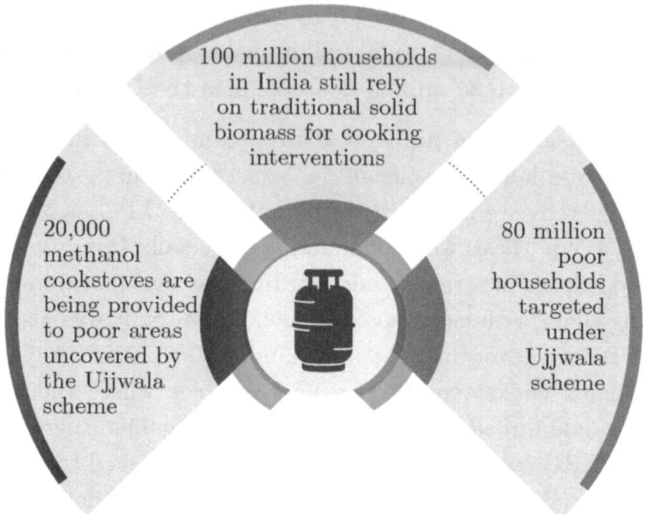


In the case of LPG, achieving the reliability of 'on demand' supply, especially in far flung areas, is likely to take time because of the aforementioned factors. As far as methanol cookstoves are concerned, in African countries where ethanol/ methanol cookstoves have been piloted, distribution at an affordable consumer price was observed to be a key determinant. Given the fact that it is proposed to source methanol from Assam for utilisation in Uttar Pradesh, the reliability and the cost of delivery of methanol becomes critical.

## Clean Cooking Solutions

One way of addressing the last mile accessibility challenge is to move to cooking solutions that are not dependent on fuels transported long-distance; for instance cooking devices run on electricity. The electricity could be from the grid or from decentralised renewables. The earlier traditional, coil type, electric devices were of course extremely inefficient but the scenario has completely changed with the introduction of electric induction cookstoves.

For a large-scale shift to electric induction cookstoves, there are certain prerequisites that need to be addressed in a concerted fashion. The first one is about strengthening the supply infrastructure even for electricity. Through concerted government efforts via schemes such as 'Pradhan Mantri Sahaj Bijli Har Ghar Yojana (SAUBHAGYA)' to ensure universal electricity access, it is not too far-fetched to imagine regular and reliable electricity supply to all rural households sooner rather than later, which then can also be used for induction cookstoves. However, for this to effectively take shape, availability of low-wattage, high-efficiency electrical cooking appliances that meet rural cooking requirements, has to be ensured.

This is followed by new products on the horizon, such as solar induction cookstoves and solar thermal cooking systems. These, in conjunction with technologically advanced 'pre-paid' or 'pay-as-you-go' features, could very well be the harbinger of a new era. With prices of solar modules on a downward trajectory, soon it may be possible to offer induction cookstoves coupled with standalone/decentralised solar systems for price-sensitive rural markets. And that will be true leap-frogging! Certain strata of society may still need a helping hand from the government to aid in the transition to clean cooking, the economic benefits of which far outweigh the financial costs.

## The Future of Cooking

While 'fuel stacking' in the context of cooking energy is conventionally used more as an indication of progression along the fuel ladder, there is no reason for the rural users to restrict themselves to one primary cooking fuel or device only. The fuel stacking then may very well be thought of as using multiple clean fuels/devices, contingent upon requirements and convenience. The transition could, therefore, be not to one, but to a combination of clean cooking solutions just like urban households that use LPG as well as electricity for cooking and use multiple electrical cooking appliances simultaneously.

*Mr Amit Kumar is Senior Fellow and Senior Director at The Energy and Resources Institute*

## Bibliography

International Energy Agency, (2017). *Energy Access Outlook 2017: From Poverty to Prosperity.* Retrieved from https://www.iea.org/publications/freepublications/publication/WEO2017SpecialReport_EnergyAccessOutlook.pdf

India's baseline trend in the food-water-energy nexus shows increased electricity consumption, greater groundwater depletion and bigger subsidy burden on state governments

# Implications of Enhanced Energy Use in the Agricultural Sector in India: Food, Water and Energy Nexus

*By Shripad Dharmadhikary, Sreekumar Nhalur, and Ashwini Dabadge*

## Understanding the Food-Water-Energy Nexus in the Agriculture Sector

Agriculture, particularly food grain production, has shown significant growth in India since independence. The total production of food grains increased close to five times from 1950-51 to 2010-11, going from 51 million tonnes (MT) to 245 MT. The increase in production can be attributed to a combination of factors like improved seeds, increased use of chemical fertilisers, better irrigation facilities, extension services, and procurement and price support.

Increased energy use has been the facilitator in many of these factors, like irrigation, fertilisers, and transport. An estimate of primary energy consumption in agriculture done by Prayas (Energy Group) revealed that for the year 2009-10, almost 40% of energy was used for fertilisers, 28% for irrigation, 15% for transport and 8% for mechanisation. In the food-water-energy nexus, irrigation is the main point of intersection. India's agriculture sector accounts for 80% of our water consumption.

Irrigation, driven by groundwater and electricity, has played a significant role in achieving the growth in agricultural production. About 70% of the paddy and wheat production in India is from irrigated areas. The fact that these two crops constitute more than 75% of the total food grain production in India underscores the significance of irrigation.

There has been a gradual shift towards groundwater in India's irrigation trend. Groundwater's share in the net irrigated area has increased from about one-third in the 1950s to almost two-thirds at present. This domination of groundwater has been achieved essentially by the extensive use of energy, mainly electricity for tube wells and open wells.

The electricity consumption in agriculture has risen from 3,465 Million Units (MU) in 1969 to 1,87,493 MU in 2016. Almost all the electricity that is used in agriculture is used for pumping. Similarly, more than 85% of the total energy used for groundwater irrigation, or more accurately, pumped irrigation comes from electricity. The total number of electric and diesel pump-sets in India has been steadily increasing and is now estimated to be around 26 million.

There are several reasons for this shift towards groundwater-based irrigation. One is that groundwater-based irrigation places much greater control in the hands of farmers, especially in terms of the timing of irrigation. Greater control allows optimal benefits of inputs like fertilisers, better seeds, etc. The productivity of groundwater irrigated crops is significantly higher. Other reasons include the difficulty of surface or canal irrigation reaching many areas, increasing costs of developing canal-based large surface irrigation systems, gaps between the potential created and utilised for many of the surface irrigation schemes, and that groundwater irrigation systems are modular in nature and farmers can put them up individually. Even in areas commanded by canals, farmers often opt for groundwater-based irrigation where part of the water drawn from the ground is the seepage of the surface schemes.

## Future Trends

The advantages of groundwater, and the need and plans to increase food production in the country indicate that the dominance of groundwater in irrigation will continue to grow, thus implying greater electricity use in agriculture.

The National Commission for Integrated Water Resources Development Report projected the food grain requirement for India at 320 million tonne in 2025, and at 494 million tonne in 2050 as compared to the current (2016-17) production of 275 million tonne. The government's 'Bringing Green Revolution to Eastern India' programme plans "to harness the water potential for enhancing rice production in eastern India which was hitherto underutilised" in seven states – Assam, Bihar, Chhattisgarh, Jharkhand, Odisha, Eastern Uttar Pradesh and West Bengal – with a focus on wheat and rice crops, and groundwater irrigation. In these states, groundwater available for future irrigation development is 98 Billion Cubic Meter (BCM), or 60% of the 162 BCM available in the country. This will also imply an equivalent increase in the electricity and energy use. The 19$^{th}$ Electric Power Survey from the CEA projects electricity use in agriculture to increase to 354 billion units in 2027, nearly double from 2016. The use of energy in irrigation could further increase because of several other factors like declining groundwater levels, which would necessitate higher energy to pump the same amount of water.

Another trend that is likely to continue is the shift from diesel to electricity as the source of pumping energy. The electricity grid reached all the villages in 2018 and it is a matter of time before most diesel-based pump sets are replaced by electric ones. But this trend may see some resistance if the electricity availability is not adequately extended or its quality of supply is not maintained.

With growing electricity consumption in agriculture, subsidy support for this is likely to increase significantly. An analysis of data reported by the Power

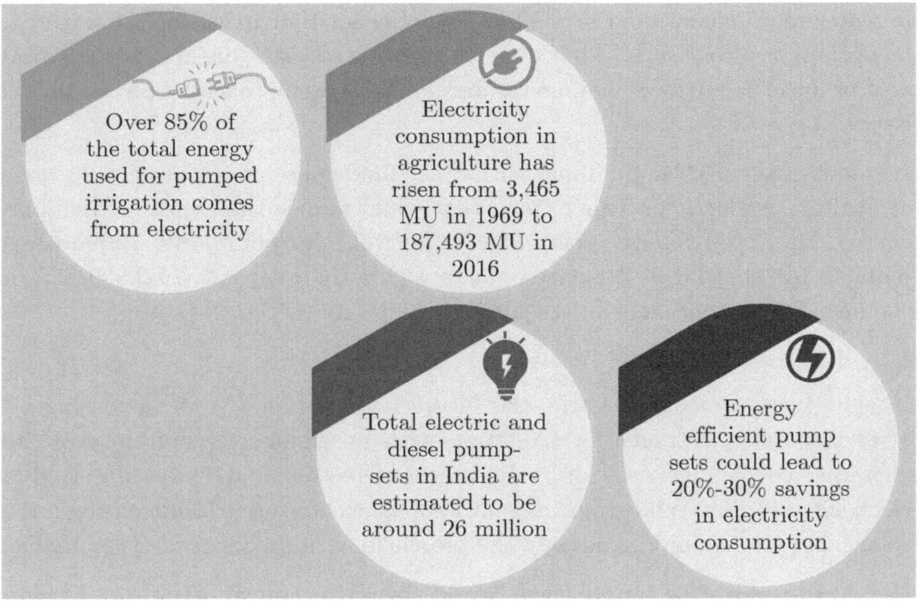


Finance Corporation shows that between 2006-07 and 2015-16, the subsidy required for agriculture in 10 states with significant electricity consumption* grew from around ₹27,900 crores to around ₹91,000 crores. The share of state governments in this quantum is higher than the cross-subsidy, which is contributed by industrial and commercial consumers. The state governments' share has also been increasing. However, the financing of this growing agricultural electricity subsidy requirement will become more challenging.

The baseline trend in the food-water-energy nexus is therefore worrisome: increased groundwater depletion in some 61% of India's wells, increased electricity consumption, and increased subsidy burden on the state governments. None of these parameters is on a sustainable trajectory.

Suggestions to shift towards crops that are more suitable to the agroclimatic character of a region, if implemented, have the potential to moderate the growth

*Andhra Pradesh, Gujarat, Haryana, Karnataka, Madhya Pradesh, Maharashtra, Punjab, Rajasthan, Tamil Nadu and Uttar Pradesh. Agricultural consumption in these states formed 97% of the total agricultural electricity consumption in the country in 2013-14.

in water requirement, and hence the energy needed for irrigation. However, to be optimistic about such a shift over the next 10-15 years, one must break away and bring in far more fundamental and innovative approaches to bear on the issue.

Increasing use of solar pumping, either as standalone pumps or through solar agriculture feeders, is another trend that could reduce the use of conventional energy. Many states have programmes to install solar pump sets; Maharashtra plans to install solar feeder-based plants with a total capacity of 2,500 MW in the next few years that will feed 20% of the pump sets in the state.

## Implications of These Trends

In a business-as-usual scenario, the future trajectory is one of increasing use of energy and electricity for irrigation with no significant shift in cropping patterns, continuing risks of groundwater depletion and increasing load of electricity subsidy. While some amelioration is expected due to increasing use of solar energy for pumping, largely the present-day practices are unsustainable.

However, innovative approaches could help in transitioning to a more sustainable and equitable scenario. Serious efforts towards less water-intensive cropping patterns will be crucial. Care should be taken that mistakes made in the green revolution phase are not repeated. Extensive soil-water conservation and rainwater harvesting schemes must also be undertaken to maintain the water balance and reduce the energy needed for pumping.

The share of solar energy in the agricultural pumping electricity will increasingly rise, with solar pumps in some states and feeders in others. Progress in development of economic storage options would provide significant support to the deployment of solar energy in agriculture. Energy efficient pump sets could lead to 20% to 30% saving in electricity consumption. Measures to improve water and land-use efficiency like organic farming, sprinkler and drip irrigation, better regulation of groundwater extraction and use, and participatory community groundwater management would also lead to efficient use of energy.

This transition is possible only when the strong interlinkages between food, water and energy are recognised and analysed in an integrated manner to present a long-term, comprehensive set of institutional reforms and measures. While conventional practices have brought us thus far, they might not take us beyond, for they are unsustainable. Therefore, it is imperative that all stakeholders join hands towards developing new institutional mechanisms that support India's ongoing efforts towards a sustainable future.

*Mr Shripad Dharmadhikary is Senior Research Fellow at Prayas (Energy Group)*
*Mr Sreekumar Nhalur is Member at Prayas (Energy Group)*
*Ms Ashwini Dabadge is Research Associate at Prayas (Energy Group)*

## Bibliography

Central Electricity Authority, MoP, GoI, (2017). *Report on 19th Electric Power Survey of India, Vol.1.* New Delhi, India.

Central Electricity Authority, MoP, GoI, (2016). *All India Electricity Statistics- General Review 2016.* New Delhi, India.

Department of Agriculture & Cooperation, MoA&FW, GoI, (2015). *Bringing Green Revolution to Eastern India: Operational Guidelines.* Retrieved from: http://rkvy.nic.in/static/download/pdf/BGREIGuidlines.pdf

Directorate of Economics & Statistics, MoA&FW, GoI, (2016). *Land Use Statistics 2013-14.* Retrieved from: http://eands.dacnet.nic.in/LUS_2013-14/Cover-Page.pdf

India Infoline News Services, (2018). *India to witness record food grain production in 2017-18, advance estimates suggest.* Retrieved from: https://www.indiainfoline.com/article/news-top-story/india-to-witness-record-food-grain-production-in-2017-18-advance-estimates-suggest-118022800062_1.html

Ministry of Agriculture & Farmers Welfare, GoI (2016). *State of Indian Agriculture 2015-16.* Retrieved from: http://eands.dacnet.nic.in/PDF/State_of_Indian_Agriculture,2015-16.pdf

Ministry of Power, GoI (2011). *Annual Report 2010-11.* New delhi, India.

Ministry of Water Resources Development, GoI, (1999). *Report of the National Commission for Integrated Water Resources Development.* New Delhi, India.

Power Finance Corporation, (2018). *The Performance of State Power Utilities for the years 2013-14 to 2015-16.*Retrieved from: http://www.pfcindia.com/Home/VS/29

Power Finance Corporation, (2016). *The Performance of State Power Utilities for the years 2012-13 to 2014-15.* Retrieved from: http://www.pfcindia.com/Home/VS/29

Power Finance Corporation, (2013). *The Performance of State Power Utilities for the years 2009-10 to 2011-12.* Retrieved from: http://www.pfcindia.com/Home/VS/29

Power Finance Corporation, (2010). *The Performance of State Power Utilities for the years 2006-07 to 2008-09.* Retrieved from: http://www.pfcindia.com/Home/VS/29

India has a vision of having at least 30% of its vehicle fleet as electric by 2030

# The Future of Mobility is Electric

*By Anirban Ghosh*

A NITI Aayog report released in 2018 said India can save 64% of anticipated passenger road-based mobility related energy demand and 37% of carbon emissions in 2030 by pursuing a shared, electric and connected mobility strategy. This, according to the report would result in a reduction of 156 million tonnes of oil equivalent (Mtoe) in diesel and petrol consumption for that year. At $52/ barrel (bbl) of crude oil, this would imply net savings of roughly ₹3.9 lakh crore (approximately $60 billion) in 2030.

But there are other reasons to go electric. A 2017 study by the European environmental non-profit Health and Environment Alliance (HEAL) said India spent $16.9 billion on oil, gas and coal subsidies in 2013 and 2014 but the health costs to meet the burden of air pollution linked diseases was eight times more at $140.7 billion. Another factor is the efficacy; according to the late Sir David JC MacKay, Cambridge physicist and government scientific adviser, "Electric vehicles can deliver transport at an energy cost of roughly 15 kilowatt hours (kWh) per 100 km. That's five times better than our baseline fossil car, and significantly better than any hybrid cars".

## Clean Mobility

Mobility when shared, connected, reliable and electric would be clean. The Government of India is spurring action by helping citizens make this transition. The Government of India has a vision of having at least 30% of the vehicle fleet as electric by 2030. Transport is an important focus area of India's climate strategy with the sector accounting for 13% of India's energy-related carbon dioxide ($CO_2$) emissions. The Nationally Determined Contributions (NDCs) as submitted under the Paris Agreement sets an economy-wide decarbonisation target of 33-35% by 2030 over 2005 levels and highlights the role of urban transport, intercity transportation infrastructure, sustainable logistics, and inland waterways to achieve these reductions.

Very often, a concern raised is that by going electric, are we shifting the emissions to the grid or actually reducing them? No doubt net emissions are lower than the traditional internal combustion engines (ICE) over the life cycle of electric vehicles (EVs), but India's ambitious renewable energy plans, which now target 227 GW by 2022, will make the grid greener. Thus, with a lower emissions factor of the grid, the net reduction in emissions would be much greater.

## Impact on Imports

Another argument that one hears is that electrification of mobility may not necessarily reduce imports – it will reduce oil imports, but increase battery imports. Evidence suggests that even in the near future, wherein India will have to depend on imports for cells while battery pack and fabrication can be done domestically, the import burden of batteries would be lower than the oil import bill. With time, while battery costs decrease, risks of raw material availability such as lithium, cobalt, and nickel among others will be key for India to watch for. India has already put in place a draft National Energy Storage Mission (NESM), which aims to strategically look at energy storage across the value chain, from battery manufacturing to end use applications.

## Creating Demand Incentives

While supply constraints may exist, India must focus on creating sufficient demand-side incentives in the short-medium term to drive the transition to electric mobility. Demand incentives can include enabling a robust ecosystem of charging infrastructure, subsidy for EV buyers, tax incentives and so on. One of the key bottlenecks is the availability and density of public charging infrastructure within a city/region. In metro cities, the opportunity cost of land is a key barrier for operators to install charging stations. There needs to be focus on enhancing asset utilisation of land, for example, in the night, after malls close, their parking lots can be leased out to fleet operators for charging their EV fleets. With the FAME (Faster Adoption & Manufacturing go Hybrid and Electric Vehicles) 2.0 subsidy programme, the government has signaled its intent of facilitating the e-mobility transition. Various states in India have also

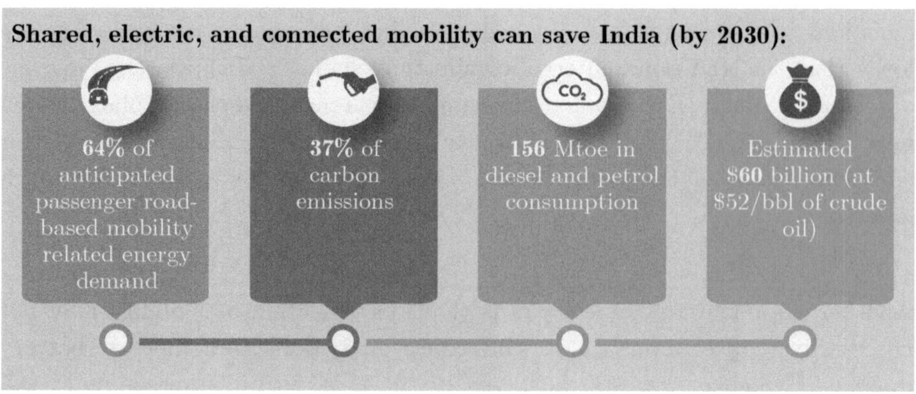

given preferential electricity tariffs for charging of EVs at designated/public charging stations, as well as waiver of road and registration tax for EV buyers.

New business models can be developed once access to EVs becomes a norm. Drivers and fleet operators of EVs could play the role of prosumers, i.e., of customers who design products for themselves and start producing them. Another opportunity is for producers and consumers of energy services such as vehicle-to-everything (V2X), which is the passing of information from a vehicle to any entity that may affect the vehicle, and vice versa through smart charging. These new energy services will create additional opportunities for revenue sharing between the vehicle owners and the energy suppliers that would reduce the total cost of ownership of EVs and accelerate their market penetration.

An example of such an exercise can be seen at the EUREF campus of Schneider on the outskirts of Berlin, where the EV charging stations are integrated in the local micro smart grid with solar and wind generation. Schneider Electric has collaborated with the Innovation Center for Mobility and Societal Change to complete a micro smart grid that features artificial intelligence and machine-to-machine learning capacity that actively optimises EV charging. Another example is the Vehicle 2 Grid (V2G), which is a system in which plug-in electric vehicles communicate with the power grid to sell demand response services by either returning electricity to the grid or providing frequency regulation services. It will certainly be an interesting proposition and could play a role in grid balancing. Closer to home, Andhra Pradesh has already made a provision for this in its FY19 electricity tariff order and the state plans to have 1 lakh electric vehicles by 2020. The Enel Group, which is one of Italy's leading energy companies, has also successfully carried out pilots on V2G applications in Europe. Second life value chain of EV batteries could also potentially disrupt the market – bringing down the capex of an EV and opening new markets for business.

## Access to Finance

Access to finance will play an important role. With the initial capex of an EV on the higher side as compared to an ICE vehicle as well as investment required for charging and parking infrastructure, availability of low-cost and patient capital will be critical. The overall policy framework in terms of an integrated transport and energy policy will go a long way in providing the necessary support to drive the e-mobility transition. India can draw lessons from the experiences of Scandinavian countries and its neighbour China.

With a growing demand for mobility in India, the potential for scaled adoption of EVs across different customer segments, including passenger movement (shared and personal) as well as freight movement is significant. The dividends of a shift to a cleaner and low-carbon transport system will address the issue of mitigating climate risks, enhanced mobility standards and higher savings for end-consumers. The future of mobility in India will need to be shared, convenient, connected, reliable and electric.

*Mr Anirban Ghosh is Chief Sustainability Officer of the Mahindra Group*

## Bibliography

MacKay, D. J. C. (2009). *Sustainable Energy – without the hot air*. Cambridge: UIT Cambridge Ltd.

India is delivering the world's largest energy efficiency portfolio, having successfully drawn wide participation from diverse sectors such as industry, buildings and residential

# **Powering the Powerhouse:** India's Energy Efficiency Leap

*By Saurabh Kumar*

When India ratified the Paris Agreement in 2016, the nation's Nationally Determined Contributions (NDCs) were considered highly ambitious by many quarters of the global climate change community. To some extent, this reaction was a fair response to the nation's socio-economic imperative of achieving manufacturing and infrastructure growth at an unprecedented scale, the energy intensity of which would contribute to record emissions.

Nonetheless, India accepted its share of the burden for global emission reductions in tandem with, and as part of, its development ambition, accelerating efforts that would enable the nation to achieve these dual objectives in a sustainable manner. Programmes targeted at scaling energy efficiency and percolating its benefits to every strata of society and economy have been especially effective in this regard.

With the initial success and effectiveness of these programmes now universally evident, it is time for stakeholders to make lasting commitments towards these proven initiatives and approaches, especially as we now power towards our goal of reducing our emissions intensity by 33% to 35% within the decade, while simultaneously striding towards a more secure energy future.

## Efficiency by Design

In recent years, India has set milestone after milestone in energy efficiency, earning its hard-won global leadership position. Today, India delivers the world's largest energy efficiency portfolio and has successfully incentivised large-scale participation by stakeholders across the spectrum, including in sectors as diverse as buildings, manufacturing, residential, and municipal.

This era of energy efficiency performance was forged by many policy initiatives that date back almost five decades. These include the establishment of the Fuel Policy Committee in the 1970s in the wake of the fuel crisis to optimise electricity generation and transmission; identification of energy saving targets and supporting of policy recommendations by the 1981 Inter-Ministerial Working Group on Energy Conservation (IMWG); and, milestone legislations such as the Energy Conservation Act 2001 and the Electricity Act 2003, which established the foundation of energy efficiency in Indian industry.

Industry's pivotal role and success in driving forward the nation's agenda for energy efficiency was realised with the implementation of the National Mission for Enhanced Energy Efficiency's (NMEEE) market-based scheme Perform Achieve Trade (PAT), which incentivises large energy-intense industries to reduce energy consumption and monetises the energy so saved with tradeable 'Energy Saving Certificates', or ESCerts. In the first phase of PAT that ended in 2014-15, India realised carbon emission reductions of 31 million tonnes between 2012-2015, translating to ₹47,185 crore in energy savings.

## Scaling the Change

With their immense cost-saving potential, energy efficient technologies even captured the interest and intrigue of the general public. Grounded in the concept of demand aggregation, the Unnat Jeevan by Affordable LEDs and Appliances for All (UJALA) programme was launched to stimulate domestic manufacture and make energy efficient lighting technologies affordable and accessible across the country. Combining demand aggregation with procurement, the programme reduced LED bulb prices to almost a tenth of previous market prices. Directly competing with their imported counterparts on metrics of quality, domestically manufactured LED bulbs, with their durability, better light output and lower energy bill appealed to the masses – and a movement was born. Today, UJALA is the world's biggest distribution programme of energy efficient LED bulbs with households, commercial establishments and municipalities saving over 41 billion kilowatt hour (kWh) of energy and over ₹16,000 crores in electricity costs.

This replicable, technology- and geography-agnostic programme has also become the foundation and launch-pad for the transformation of markets for several other energy efficient technologies, albeit on a smaller scale. Energy Efficiency Services Limited (EESL), the chief architect of UJALA, has endeavoured to adopt this approach for agriculture and municipal pump sets, street lighting, smart meters, electric mobility, and solar lamps across India and the world. EESL's approach of aggregating demand ensures that the market attracts industry participation, while passing the benefits of cost reduction – created through bulk procurement – on to energy consumers.

The combined effect of India's energy efficiency and conservation measures has been a 58% decline in the national economy's energy intensity between 2005-06 and 2015-16, projected to decline further by 37% in the next 20 years.

Today, India's energy efficiency programmes invite global interest from countries seeking to replicate them. There is also regular funding from multilateral bodies to scale such innovative programmes even further. With its population and emerging economic might, India has the potential to not only become a global powerhouse, but also one that is sustainable by design. And the solutions that work in India – a challenging, highly complex market – would definitely work in many other geographies.

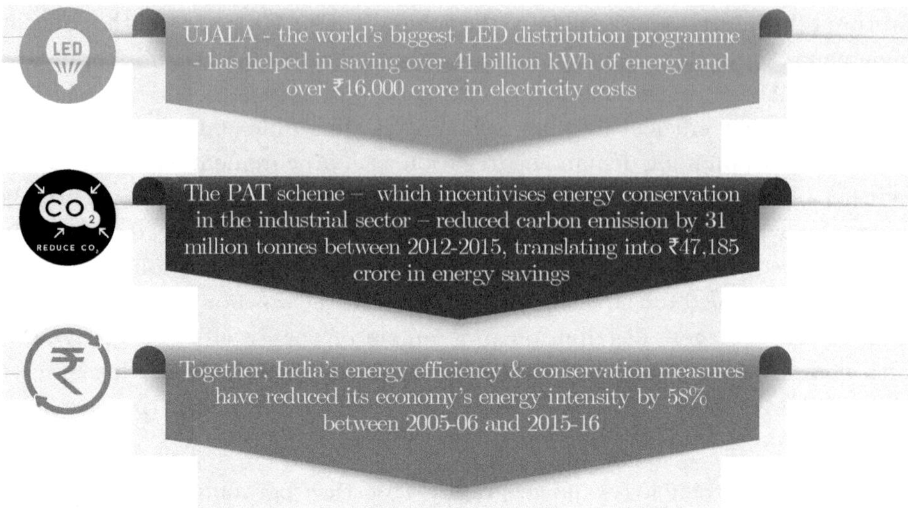


## Assimilating Efficiency

The ESCert trading and EESL's multi-focused, international programmes have already provided compelling evidence that, whether through market mechanisms or regulatory action, energy efficiency directly translates not only to monetary savings, but also to holistic economic development. With technology advancements and innovations, the initial quantum of, and payback on, investments in energy efficiency are changing to the investor's favour, while delivering extensive social and environmental benefits.

Thus able to demonstrate its capabilities in executing tangible actions and achieving measurable results in the global movement towards climate change mitigation, India has now reached a juncture for assimilating efficiency as a permanent facet of its development process. In order for us to accelerate our progress along the path of self-sustaining development, Indian stakeholders, industry and government included, need to integrate metrics of energy efficiency in the broader investment decision making process.

To some extent, this process has already begun. The foremost example is the smart meter, a technology which promises to cut India's energy distribution losses, while augmenting efficiencies in energy-usage by consumers. The smart meter will also support renewable energy integration and net metering, which in turn will be vital to enable next generation e-vehicle owners to affordably fuel their mobility needs. India envisions bringing 6 million electric and hybrid vehicles on India's roads by 2020, targeting savings of 120 million barrels of oil and 4 million tonnes of $CO_2$.

There are also international opportunities being explored for trigeneration, which uses natural gas or biogas to simultaneously support cooling, heating and power generation needs. Manufacturing units, hotels, hospitals, and airports across India have already evinced interest in such technology that endeavours to optimise efficiencies in the power generation process while harnessing waste heat and byproducts to support industrial activity.

But, there is abundant potential yet to be tapped. With new avenues for industrial growth emerging on a daily basis as the nation powers towards its gross domestic product (GDP) growth targets, the opportunities for harnessing energy efficiency extend beyond technology development. Marching in momentum with the world's energy transition and its own sustainable growth aspirations, India is progressively finding opportunities to make energy efficiency a prominent and foremost consideration in the energy consumer's conscience.

What India has demonstrated is that technology deployment cannot be a standalone effort, but part of a vision that focuses on coherently transforming the energy value chain to become both sustainable and accessible for all. Technologies are just the starting point for energy transitions; results come from business models tailor-made for the Indian context, but also driven by increasingly aware and knowledgeable consumers. While India's current and future energy transition infrastructure has been tested for proof of concept globally, they will only matter if they 'work for India' – with a buy-in based on clearly visible incentives and investment returns.

## Beyond Savings

While India needs to bring 300 million people to the grid and create large-scale economic opportunities, our Paris Agreement goal of reducing emission intensity has allowed India to make interventions right at the heart of energy -usage – bringing energy efficiency to the fore.

But, if India is to lead the world's energy transition, as it is doubtless capable of doing, the country needs to harness sustainability-focused economic transformation in a far more rigorous manner, and escalate the discussion on, and consideration of, energy efficiency beyond technology deployment and cost savings. Energy efficiency and responsible development considerations need to become a primary driver of the investment and business growth process.

*Mr Saurabh Kumar is Managing Director of Energy Efficiency Services Ltd.*

---

## Bibliography

British Petroleum, (2018). *BP Energy Outlook 2018: Country and regional insights - India.* Retrieved from: https://www.bp.com/content/dam/bp/en/corporate/pdf/energy-economics/energy-outlook/bp-energy-outlook-2018-country-insight-india.pdf

Ministry of Statistics and Programme Implementation, GoI, (2017). *Energy Statistics 2017.* Retrieved from: http://www.mospi.nic.in/sites/default/files/publication_reports/Energy_Statistics_2017r.pdf.pdf

The Economic Times, (2017). *Over 400 industries reduced $CO_2$ emission by 2% in 2012-15.* Retrieved from: https://economictimes.indiatimes.com/news/company/corporate-trends/over-400-industries-reduced-co2-emission-by-2-in-2012-15/articleshow/61219834.cms

The Washington Post, (2018). *Every village in India now has electricity. But millions still live in darkness.* Retrieved from: https://www.washingtonpost.com/world/asia_pacific/every-village-in-india-now-has-electricity-but-millions-still-live-in-darkness/2018/04/30/367c1e08-4b1f-11e8-8082-105a446d19b8_story.html?noredirect=on&utm_term=.b8765d4ea6e5

Emissions from heavy industry and long distance road transport – sectors which are not easily or cost effectively electrified – will be 'hard to abate' but not impossible

# Tackling the Hard-To-Abate Sectors in India's Energy Transition

*By Nitin Prasad*

India has made great progress in tackling the challenges of climate change. It has invested in renewable energy, promoted electric vehicles and biofuels, and pledged to reduce its emissions intensity – greenhouse gas (GHG) emissions per unit of gross domestic product (GDP) – by 33% to 35% below 2005 levels by the year 2030. However, India also needs to consider how it manages its 'hard-to-abate' sectors which are not easily or cost-effectively electrified, and where it is therefore harder to reduce emissions. These include heavy industries that produce iron and steel and cement as well as long-distance road transport, shipping, and the fast-growing aviation business. These sectors, fundamental to the growth of a large emerging economy such as India, will continue to rely on fossil fuels to provide extremely high temperatures, chemical reactions or dense energy storage for many decades. This means they will continue to produce GHG emissions.

Take for instance, industry, which accounts for 25% of the country's total GHG emissions. Heavy industry – namely iron and steel and cement – makes up nearly 67% of this figure, while the rest comes from chemicals and fertilizers, non-ferrous metals, and textiles. When it comes to transport, battery-electric and hydrogen fuel cell vehicles could make up 80% of the world's passenger cars over the coming decades, according to Shell scenarios.* However, electrification will be a challenge for the aviation, maritime, and long-distance trucking sectors which require dense energy carriers. It will be hard, but not impossible, to reduce emissions from these sectors.

## Public-Private Partnerships to Promote Energy Efficiency

Governments and businesses will need to work together to make it happen. Governments, through their policies, can provide a level-playing field for the adoption of lower carbon fuels through measures, such as carbon pricing mechanisms. They can also develop key infrastructure and support the development of new technologies and energy efficient processes. The private sector could then become the engine for growing, deploying, and integrating new technologies, and introducing customers to lower carbon options. Both need to work together to promote energy efficiency, and increase the adoption of lower carbon fuels and conversion technologies.

## Moving to Natural Gas

Coal, which makes up 44% of India's total primary energy consumption, is the primary source of emissions from industry. Switching from coal to natural gas, the cleanest burning fossil fuel, is one of the easiest ways to reduce emissions. It emits between 45% and 55% less GHG emissions than coal when burnt

*Shell Scenarios are not intended to be predictions of likely future events or outcomes and investors should not rely on them when making an investment decision regarding Royal Dutch Shell plc securities. Please read the full cautionary note on www.shell.com/scenarios/sky.

to generate electricity, according to the International Energy Agency (IEA) data, and less than one-tenth of the air pollutants to fuel transport, heat and light homes, and power industries. At the same time, wider use of government schemes like the PAT (Performance Achieve Trade), a market-based trading mechanism, can encourage industries to be more energy efficient. In the long-term, industries can also adopt new production processes, on the lines of reducing, reusing, and recycling materials. In addition, they can consider Carbon Capture, Usage and Storage (CCUS) technologies which capture up to 90% of carbon dioxide ($CO_2$) emitted in energy-intensive industries and either channel it for other purposes or store it safely deep underground. On this point, India has already said it would increase its forest cover to create an additional carbon sink of 2.5 to 3 billion tonnes of $CO_2$ equivalent by 2030.

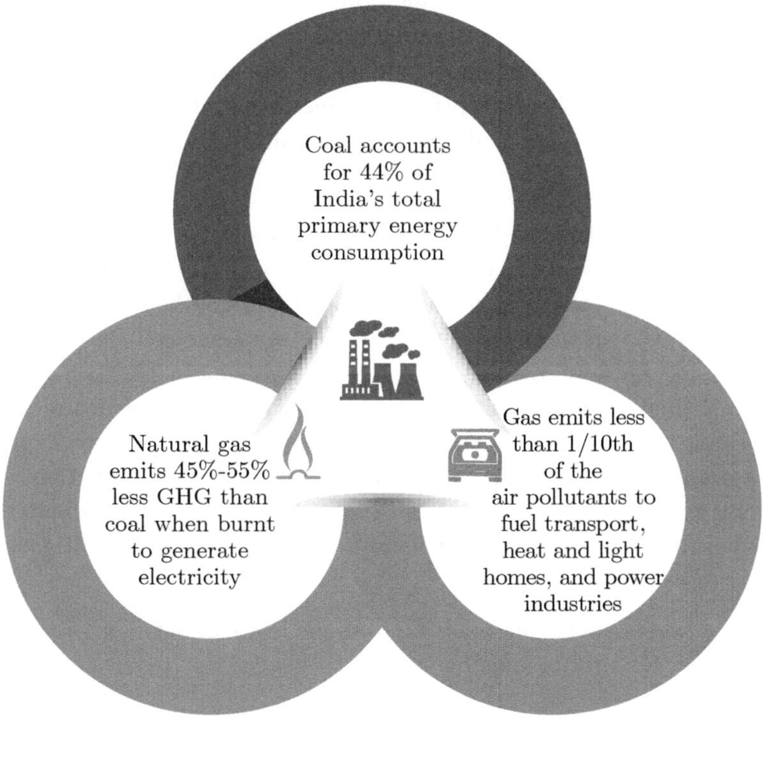

## Decarbonising Aviation

When it comes to heavy transport, the aviation and maritime sectors produce relatively fewer GHG emissions than trucking. These, however, are growing fast. India's aviation sector is growing at above 10% annually and the maritime sector's consumption of energy is likely to go up when India implements its Sagarmala inland waterways project. This will serve to modernise India's ports and create new economic opportunities by connecting its coastline and strategic maritime routes with road, rail, and air passages. As this plan is still in its infancy, there is a window of opportunity to encourage the use of cleaner fuels in this sector. In addition, the International Maritime Organization's recent ruling to impose a 0.5% sulphur fuel cap by 2020 might also encourage the adoption globally of liquefied natural gas (LNG) as a compliant fuel that also emits less carbon. When it comes to aviation, innovative and scalable solutions that go beyond bio-jet fuel are required; this may take decades. Therefore, in the near term, India would need to work on decarbonising the industry's associated businesses, such as airport operations. It would also need to consider nature-based solutions, such as afforestation, to offset emissions.

## India's Future Energy Needs

In India, trucking makes up 10% of total petroleum product consumption. Switching to cleaner fuels will go a long way towards emission reductions. In the short term, since electrification is not feasible for long-haul heavy trucking, opting for LNG, the cleanest burning fossil fuel, is one option. In the medium to long term, as the development of waste-based advanced bio-fuels advances and matures, they can be used, not only in trucks, but also in aircraft and ships. Enabling solutions across these key sectors will become more important as India's economy grows. This is because from now until 2040, the world's greatest energy demand will come from India. It will take a coordinated and collaborative approach to meet the country's energy needs while mitigating its carbon emissions at the same time.

The path to a low-carbon economy is not a single big leap but many progressive steps that everyone – India and the world – needs to walk together.

*Mr Nitin Prasad is Chairman of Shell Companies in India*

## Bibliography

Bery, S., Ghosh, A., Mathur, R., Basu, S., Ganesan, K. & Owen-Jones, R., (2017). *Energizing India: Towards a resilient and equitable energy system.* New Delhi, India: SAGE Publications Pvt. Ltd.

Council on Energy, Environment and Water, (2017). *Greenhouse Gases Emission Estimates for India – Industry Subsector (Time-series: 2007 to 2012).* Retrieved from: http://www.ghgplatform-india.org/Images/Publications/PDF%201_Green%20House%20Gases%20Emission%20Estimates%20For%20India%20-%20Industry%20 Subsector.pdf

GHG Platform India International Energy Agency, (2017). *World Energy Outlook 2017.* Retrieved from https://www.iea.org/weo2017/

Natural Resources Defense Council, (2017). *The Road From Paris: India's Progress Towards its Climate Pledge.* Retrieved from: https://www.nrdc.org/sites/default/files/paris-climate-conference-India-IB.pdf

Shell International BV, (2016). *A Better Life with a Healthy Planet: Pathways to net-zero emissions. A new lens scenario supplement.* Retrieved from: https://www.shell.com/energy-and-innovation/the-energy-future/scenarios/a-better-life-with-a-healthy-planet/_jcr_content/par/textimage_494361683.stream/1475857583070/d58a157055bd08857b92a9fcf7abde59277f6484730ad073ec37524e6f9d092f/scenarios-nze-brochure-local-print-awv9.pdf

Singh, A., (2016). A study of Current Scenario of Aviation Sector in India. In *International Journal of Innovative Knowledge Concepts, Vol. 2, Issue 4.* Retrieved from: https://www.researchgate.net/publication/302261935_A_study_of_Current_Scenario_of_Aviation_Sector_in_India

## About the Editors

**Dr Ajay Mathur** is Director General of The Energy and Resources Institute, New Delhi, and a member of the Prime Minister's Council on Climate Change. He co-chairs the Energy Transitions Commission. He was Director General of the Bureau of Energy Efficiency in the Government of India from 2006 till February 2016, and responsible for bringing energy efficiency into Indian homes, offices and factories, through initiatives such as the star labelling programme for appliances, the Energy Conservation Building Code, and the Perform, Achieve and Trade programme for energy-intensive industries. Dr Mathur was earlier with TERI; headed the Climate Change Team of World Bank in Washington DC; was President of Suzlon Energy Limited; and also headed the interim Secretariat of the Green Climate Fund.

**Lord Adair Turner** chairs the Energy Transitions Commission, a global coalition of major power and industrial companies, investors, environmental NGOs and experts working out achievable pathways to limit global warming to well below 2°C by 2040 while stimulating economic development and social progress. He also chairs the Institute for New Economic Thinking, and was more recently appointed Chairman of Chubb Europe. He is a non executive director of Prudential plc., and a Trustee at the British Museum. In December 2018 he joined the Advisory Board of Envision Energy, a Shanghai-based group focussed on renewable energy, batteries and digital systems. From 2008-2013, Lord Turner chaired the UK's Financial Services Authority, and played a leading role in the post crisis redesign of global banking and shadow banking regulation.

**Ms Noëmie Leprince-Ringuet** is an international expert on climate and energy transitions with the French agency for technical cooperation, currently based at The Energy and Resources Institute, New Delhi. She was previously with the French Ministry of Foreign Affairs posted in India focusing on Franco-Indian policy and diplomacy in the lead up to COP 21, and on strengthening cooperation between France and India in the fields of climate change, energy and environment. From 2010 to 2014 she was with the Technical Support Unit of the Synthesis Report of the Intergovernmental Panel on Climate Change Fifth Assessment Report. She has also worked at the Delegation of the European Union to India, Nepal and Bhutan (2010), and at the International Chamber of Commerce Headquarters in Paris (2009).

## About the Contributors

**Ms Ashwini Dabadge** is a researcher with Prayas (Energy Group), an organisation engaged in research, advocacy and capacity building in the energy sector and based in Pune, India. Her work at Prayas revolves around the developmental and socio-economic aspects of energy, particularly in the areas of cooking energy access, energy use in irrigation and modelling residential energy consumption. She has a Bachelor's in Economics from the University of Pune, a Master's in Economics from University College London and has completed a study on inflation and household expenditure under the 'Young Scholars Scheme' with the Reserve Bank of India.

**Mr Shripad Dharmadhikary** works on analysis and advocacy of water and energy issues. He is the coordinator of Manthan Adhyayan Kendra, a centre focussing on water policy and programmes from the perspective of just, equitable and sustainable development. He also works part time with Prayas (Energy Group) as Senior Research Fellow. Shripad received his Bachelor's degree in Mechanical Engineering from the Indian Institute of Technology (IIT) Bombay in 1985, and was working for 12 years in the Narmada Bachao Andolan, a mass organisation of people affected by large dams in the Narmada river basin.

**Mr Anirban Ghosh** leads Sustainability at the $20.7 billion Mahindra Group. Under his leadership Mahindra has developed an award winning Sustainability framework, become a founding member of the Carbon Pricing Leadership Coalition and the first to commit to doubling energy productivity. He has, in partnership with the World Bank, facilitated the creation of the Sustainable Housing Leadership Consortium to accelerate the spread of green buildings in India. Anirban Ghosh has been an invited speaker at Global Climate Action Summit, COP 21, The Climate Week, World Circular Economy Forum, GRI Global, EE Global, etc., featured in The Climate Reality Project and acknowledged as a Distinguished Sustainability Officer.

**Mr Manoj Kohli** is Executive Chairman of SB Energy – SoftBank Group. Before this, Manoj was Managing Director, Bharti Enterprises and responsible for the growth and operations of Bharti group of companies. Manoj's key contribution has been building Airtel as no 3 telco in the world. He was Managing Director and CEO (International), Bharti Airtel, and was responsible for leading the Africa operations, which was acquired in 2010. Manoj joined Bharti Airtel in 2002 and led Bharti Airtel's India operations for 8 years during which the customer base grew from 2 to 150 million. He was adjudged "Telecom Man of the Year" in 2000 and "Telecom Person of the Year" in 2004 by Media TransAsia and Voice & Data respectively.

**Mr Amit Kumar** is currently Senior Fellow and Senior Director of the Social Transformation and Knowledge Resource Centre at The Energy and Resources Institute, (TERI), New Delhi. In this role, he is responsible for initiatives focusing on energy access, holistic rural development, community engagement, and knowledge management. Earlier, Mr Kumar also led research activities in the fields of renewable energy and resource efficient process technology applications at TERI. He was actively involved in taking forward South-South cooperation in Africa and the Pacific Island countries, in his capacity as the Regional Programme Advisor for the Renewable Energy and Energy Efficiency Partnership. He also coordinated one of the global capacity building hubs of the UN supported Sustainable Energy for All initiative (SE4All).

**Mr Saurabh Kumar** is Managing Director at Energy Efficiency Services Limited (EESL), a joint venture of public-sector undertakings under the Ministry of Power, Government of India. Under Mr Kumar's leadership, EESL has implemented the globally-lauded Unnat Jyoti by Affordable LEDs for All (UJALA) programme and Street Light National Programme (SLNP).
Mr Kumar and his team have also been successfully implementing several

Demand-Side Measures in various sectors such as municipal functions, agriculture, and public buildings. Under his leadership, EESL is working on transformative initiatives like e-mobility, smart meters, trigeneration and decentralised solar power plants. Mr Kumar is an internationally recognised authority on the design and implementation of scalable, replicable and impactful business models for promoting energy efficiency across sectors.

**Mr Sunil Mathur** is Managing Director and Chief Executive Officer of Siemens Ltd. He has been with Siemens for over 26 years, holding several senior management positions in Germany and the UK in the Energy and Industry sectors. Starting his career in the Internal Audit team in Delhi in 1987, he moved on to join the Business Administration Group in the Energy Sector. He was Cluster CFO for South Asia in July 2008 and was Executive Director and Chief Financial Officer of Siemens Ltd. from December 2008 till December 2013. During his stint as the CFO of Siemens India, Mr Mathur was part of Siemens AG CFO's Management Team that consisted of the Sector CFOs, the Heads of the Corporate Departments and selected Country CFOs.

**Mr Rakesh Nath** is former Technical Member of the Appellate Tribunal for Electricity and former Chairperson of the Central Electricity Authority (CEA), Ministry of Power. He has about 37 years of varied experience in power sector planning, operation & maintenance of thermal and hydropower stations and transmission systems, regulation of water supply from multi-purpose hydro projects, power system operations, power trading, and electricity regulatory matters. As Chairman CEA, he worked extensively for accelerated capacity addition during the 11$^{th}$ Five Year Plan and initiated advance action for the 12$^{th}$ Plan. He was instrumental in the preparation of a proposal for a low-carbon growth strategy for the power sector till 2022. He was also an ex-officio Member of the Central Electricity Regulatory Commission, part time

Director of the Nuclear Power Corporation, and Chairman of the Bhakra Beas Management Board.

**Mr Sreekumar Nhalur** has been working in the power sector in the areas of policy analysis and Information Technology applications since 1984. After 14 years of work in the industry, in 2000 he joined Prayas, a voluntary organisation based in Pune, India, and is currently Member, Prayas (Energy Group). The Energy Group works on policy analysis, governance and civil society capacity building in the energy sector. Sreekumar is a member of government committees and is associated with several voluntary organisations. He has authored many articles, papers and booklets, including a Citizens' Primer on the electricity sector and a book on electricity sector reforms. Sreekumar received his Bachelor's degree in Electrical Engineering from IIT Bombay in 1984 and Master's degree in Power Systems Engineering from IIT Kharagpur.

**Mr Nitin Prasad** is Chairman of Shell Companies in India. He is a dynamic and accomplished business leader with 20+ years of cross-cultural professional experience, spanning industries from technology to energy, and geographies from USA, Singapore to India. Nitin, a Fortune 40 under 40 awardee, has a proven track record of managing large diverse global teams, developing/implementing strategy, innovating and leading change across cross-functional and multi-jurisdictional roles. He is passionate about Shell's purpose of powering progress together for better and cleaner energy and believes collaborations and partnerships are key to solving country-wide growth challenges. Prior to his current role, Nitin was the Managing Director for Shell Lubricants for the cluster of India, Sri Lanka and Bangladesh, which combined is the 3rd largest lubricants market in the world.

**Mr Rajani Ranjan Rashmi** is currently Distinguished Fellow at The Energy and Resources Institute, New Delhi. He was a senior member of the Indian Administrative Service (batch of 1983), and Special Secretary in the Ministry of Environment, Forest & Climate Change (MoEFCC), Government of India, from May 2016 to June 2017. In this role, he held the position of Chief negotiator for the multilateral climate negotiations and the Montreal Protocol, as well as being responsible for the areas of climate change, pollution control, and environmental impact assessment in key sectors. Previously, He held the position of Chief Secretary, Government of Manipur, from July 2017 to March 2018, and Joint Secretary in MoEFCC from 2008 to 2013.

**Mr Hemant Sahai** is the Founding Partner of HSA Advocates. For the last three decades, Hemant has been a trusted legal counsel to some of the largest corporates in India and overseas, and is widely recognised for his role in shaping the Indian legal industry. In addition he has been an adviser to PSUs, regulatory authorities, multilateral institutions, banks and financial institutions, etc. on a range of policy and regulatory issues. Hemant has served as adviser to several top government bodies/institutions, including certain extra ministerial policy advisory bodies set up by the Prime Minister's Office, the Ministry of Power, and the Ministry of New and Renewable Energy. He has served clients across industries and sectors including automotive, aviation, energy, renewable energy, and environment.

**Ms Rachika A Sahay** is Partner with HSA Advocates and part of the firm's Projects, Energy and Infrastructure and Corporate Commercial practices. She has over a decade of experience in handling a broad range of deals in the energy sector. She also has diverse experience in handling corporate transactions including domestic as well as cross-border mergers and acquisitions, private equity investments and joint ventures. Prior to joining HSA, she was with Ostro Energy Private Limited (Renewable energy platform for Actis Fund III) as Head Legal of the Group. In three years of being at Ostro, she successfully handled renewable energy projects involving an aggregate capacity 1,100 MW. She was awarded with the Rising Star award by Legal Era in 2017.

**Mr A K Saxena** is Senior Fellow and Director of the Electricity & Fuels Division at The Energy and Resources Institute (TERI), New Delhi. He has 37 years of experience in the power sector in India. During the course of his career, he was Chief Engineer at the Central Electricity Authority (CEA), Ministry of Power (MoP) until 2016. Prior to that held important portfolios as Chief (Engineering) at the Central Electricity Regulatory Commission (CERC) from 2013 to 2016, Director (Operation Monitoring) and Director (Transmission) in MoP (2005 to 2012). At TERI he manages the work of the Energy Transitions Commission India on energy transitions in the Indian power sector to develop a roadmap for low-carbon pathways.

**Mr Ajay Shankar** is Distinguished Fellow at The Energy and Resources Institute, New Delhi. He has rich and varied experience in public service, with over 40 years in the Indian Administrative Service. He served in the Government of India in key policy making positions in the areas of industry and energy. He was Secretary of the Department of Industrial Policy and Promotion, Government of India, as well as Member Secretary of the National Manufacturing Competitiveness Council. Earlier, he worked in the Ministry of Power and held the posts of Special Secretary, Additional Secretary and Joint Secretary. He played a key role in piloting the Electricity Act 2003, and the major programme for completing the universalisation of electricity access in India launched in 2005.

**Mr Sumant Sinha** is a leading first generation entrepreneur. He is Chairman and Managing Director of ReNew Power - India's largest clean energy company. After being in leadership roles with large global corporations, Sumant quit a very successful corporate career to established ReNew Power in 2011. Since then, ReNew Power has grown significantly to become India's largest renewable power producer with more than 6 GW capacity, spanning more than 100 sites across India. Sumant speaks extensively at global platforms, such as the World Economic Forum at Davos and forums organized by FT, Goldman Sachs, Fortune, Boao, Observer Research Foundation and top global universities. He has authored more than 100 articles and has been featured in leading global and Indian publications.

**Mr Thomas Spencer** is Fellow at The Energy and Resources Institute, New Delhi, where he works on decarbonisation of the Indian power sector. In this role, he coordinates the work programme of the Energy Transitions Commission India, specifically on modelling and forecasting demand of electricity in India until 2030, and on projections on the supply side to meet the demand till 2030 while transitioning to a higher share of renewables in the Indian grid. Prior to that, he worked at IDDRI between 2011 and 2017, where he was Director of the Energy and Climate Change Program between 2013 and 2016. Thomas has co-authored influential publications on the ambition mechanism, transparency arrangements and financial negotiations of the Paris Agreement.

**Lord Adair Turner** chairs the Energy Transitions Commission, a global coalition of major power and industrial companies, investors, environmental NGOs and experts working out achievable pathways to limit global warming to well below 2°C by 2040 while stimulating economic development and social progress. He also chairs the Institute for New Economic Thinking, and was more recently appointed Chairman of Chubb Europe. He is a non executive director of Prudential plc., and a Trustee at the British Museum. In December 2018 he joined the Advisory Board of Envision Energy, a Shanghai-based group focussed on renewable energy, batteries and digital systems. From 2008-2013, Lord Turner chaired the UK's Financial Services Authority, and played a leading role in the post crisis redesign of global banking and shadow banking regulation.

The Energy and Resources Institute (TERI) is an independent non-profit organisation, with capabilities in research, policy, consultancy and implementation. TERI has multi-disciplinary expertise in the areas of energy, environment, climate change, resources, and sustainability. With the vision of creating innovative solutions for a sustainable future, TERI's mission is to usher in transitions to a cleaner and more sustainable future through the conservation and efficient use of the earth's resources and develop innovative ways of minimising waste and reusing resources.

The India spin-off of the Energy Transitions Commission (ETC India) was launched in New Delhi in February 2018. ETC India is a unique, high-level, multi-stakeholder platform on energy and electricity sector transitions in India. ETC India's work programme is centred around policy analysis and recommendations, research, and outreach activities on decarbonising the Indian power sector. ETC India is led by TERI where the Secretariat is hosted, in partnership with The National Renewable Energy Laboratory of the US (NREL) and Climate Policy Initiative (CPI).

For Product Safety Concerns and Information please contact our EU representative GPSR@taylorandfrancis.com Taylor & Francis Verlag GmbH, Kaufingerstraße 24, 80331 München, Germany

**Batch number: 10725843**

Printed by Printforce, the Netherlands